Sloshing of Cyrogenic Liquids in a Cylindrical Tank under normal Gravity Conditions

SLOSHING OF CRYOGENIC LIQUIDS IN A CYLINDRICAL TANK

UNDER NORMAL GRAVITY CONDITIONS

Vom Fachbereich Produktionstechnik

der

UNIVERSITÄT BREMEN

zur Erlangung des Grades
Doktor-Ingenieur
genehmigte

Dissertation

von

Dipl.-Ing. Tim Arndt

Gutachter:

Prof. Dr.-Ing. Michael Dreyer

Prof. Dr. Emil Hopfinger
(LEGI, Universite Joseph Fourier, Grenoble)

Tag der mündlichen Prüfung: 26.10.2011

Bibliografische Information der Deutschen Nationalbibliothek

Die Deutsche Nationalbibliothek verzeichnet diese Publikation in der Deutschen Nationalbibliografie; detaillierte bibliografische Daten sind im Internet über http://dnb.d-nb.de abrufbar.

1. Aufl. - Göttingen: Cuvillier, 2012

Zugl.: Bremen, Univ., Diss., 2011

978-3-95404-124-4

Abstract

Particularly in the upper stage development of rockets (launchers), gravity dominated fluid motion in upper stage tanks (sloshing) during flight represents an undesired dynamic effect. On the one hand the sloshing forces lead to disturbances, which have to be compensated by the reaction control system. On the other hand, when cryogenic fluids are considered, the fluctuations in tank pressure may be critical under some circumstances compromising the structural stability of the tank. In this field, the utilization of cryogenic propellants represents a high challenge to layout and design of the propulsion components including the propellant tanks.

This work deals with two effects that are directly coupled to the sloshing content inside the propellant tank. To investigate these effects a dedicated test setup has been developed. At first, the damping characteristics of sloshing cryogenic nitrogen - which is used as a substitute for the rocket propellants liquid hydrogen and liquid oxygen - are determined. The results are correlated to the theory based on storable propellants. The main part of this work is linked to a characteristic pressure drop inside the propellant tank caused by the sloshing liquid. For the effect to occur, the tank must be pressurized to enable the formation of a thermal stratification below the liquid surface. Sloshing leads to the mixing of liquid in this region with subcooled liquid from the bulk. This affects the decrease of the temperature at the free surface leading to the condensation of superheated vapor. Thus, the pressure in the tank must decreases. Three different pressurization concepts are introduced in this work; self-pressurization where the tank is pressurized by evaporating liquid caused by the heat flowing into the tank. Furthermore, the tank is pressurized with gaseous nitrogen taken from an external gas bottle and at last gaseous helium from an external supply is used for pressurization purpose. By the application of helium as non-condensable gas, a significant reduction of the pressure drop is expected. The experimental results confirm this working hypothesis and therefore support the theoretical considerations described by an approach of DAS & HOPFINGER.

All results are presented in a non-dimensional form to allow the comparison to data from the literature. Furthermore, the upscaling of the current results enables the prediction for future cryogenic upper stages such as the ESC-B for the European space launcher Ariane 5.

Zusammenfassung

Schwerkraftdominierte Flüssigkeitsbewegungen in geschlossenen Tanks beschreiben im Bereich der Raumfahrt - spezieller im Bereich der Oberstufenentwicklung - einen dynamischen Zustand, welcher meist unerwünscht zu einer Störung der Lageregelung führt und unter ganz besonderen Umständen zu einem Risiko der Tankintegrität werden kann. Die Verwendung von kryogenen Treibstoffen stellt eine große Herausforderung an die Auslegung und Entwicklung der Antriebskomponenten dar, zu denen auch die Treibstofftanks gehören.

Diese Arbeit beschäftigt sich mit zwei Effekten, welche im direkten Zusammenhang mit der schwappenden Flüssigkeitsbewegung im Tank stehen. Zunächst wird das Dämpfungsvermögen von kryogenem Stickstoff - als Ersatz für die Oberstufentreibstoffe Flüssigwasserstoff und Flüssigsauerstoff - bestimmt und mit Korrelationen aus dem Bereich der lagerfähigen Treibstoffe aus der Literatur verglichen. Der Hauptteil dieser Arbeit besteht aus der Untersuchung eines charakteristischen Druckabfalls, welcher durch das Schwappen der kryogenen Flüssigkeit (Stickstoff) hervorgerufen wird. Voraussetzung für diesen Effekt ist ein bedrückter Tank, welches die Ausbildung eines Temperaturgradienten im Bereich der Flüssigkeitsoberfläche impliziert. Die Flüssigkeit wird durch die Schwappbewegung durchmischt und führt folglich zu einem Absinken der Sättigungstemperatur, zu Kondensation und zu dem Druckabfall. Es werden drei verschiedene Bedrückungskonzepte untersucht, welche im Bereich der Oberstufenentwicklung Anwendung finden könnten. Als erstes wird der Tank selbstbedrückt durch die Verdampfung, die durch die in den Tank fließende Wärme hervorgerufen wird. Als zweites wird der Tank mit gasförmigem Stickstoff bedrückt und als drittes mit gasförmigem Helium. Durch den Einsatz von Helium als nicht-kondensierbares Inertgas wird eine Reduzierung des Effektes erwartet.

Die erzielten Ergnisse in dieser Arbeit werden dimensionslos präsentiert und ermöglichen einen Vergleich mit Ergebnissen aus der Literatur und einen Ausblick auf den Umfang dieser Effekte skaliert auf die geplante kryogene Oberstufe ESC-B der europäischen Trägerrakete Ariane 5.

Acknowledgment

This work evolved from my employment as a research associate in the Multiphase Flow Group at the Center of Applied Space Technology and Microgravity (ZARM) at the University of Bremen. First of all, I would like to gratefully acknowledge Prof. Dr.-Ing. Michael Dreyer, who not only provided very helpful guidance, supervision and scientific direction, but also allocated the necessary funding for my project. Without his appreciated and valued support, this work would not exist. Especially, I would like to thank Prof. Emil Hopfinger for agreeing to supervise my work and for his participation in the examination process.

I also like to acknowledge Prof. Dr.-Ing. Hans Rath and Prof. Dr.-Ing. Reinhold Kienzler for taking their valuable time to be part of my examination committee.

At this point, I would like to honor the help of Dr.-Ing. Aleksander Grah and Dr.-Ing. Nicolas Fries to review this work and to support me by many discussions. This also includes our project partner from EADS Astrium Dr.-Ing. Kei Philipp Behruzi, Mike Winter and Dennis Haake to support me with the test facility as well as with all the important informations concerning the Ariane launcher.

Special thanks are due to my colleagues who helped to design and build the experimental hardware. Without the valued support of Frank Ciecior, Holger Faust, Ronald Mairose and Peter Prengel the experiments could have never been conducted. Stephanie Dackow did a great job organizing literature and helping with the completion of administrative duties.

Many thanks also to my colleagues from the Multiphase Flow Group Dr.-Ing. Michael Conrath, Eckart Fuhrmann, Diana Gaulke, Dr.-Ing. Jörg Klatte, Nicolai Kulev, René Sionkiewicz and Yvonne Winkelmann who worked on projects with me at ZARM.

I would also like to especially acknowledge the support of my (former) student Sven Kastens who helped with building the experimental hardware as well as supported to program the data acquisition software.

Many thanks also to Prof. Richard M. Lueptow from Northwestern University, Evanston, IL, USA, for supporting me in summer 2009 to finish my PhD thesis.

The funding of the research project by the German Federal Ministry of Economics and Technology (BMWi) through the German Aerospace Center (DLR) under grant number 50 RL 0741 is gratefully acknowledged.

Last but not least it is my pleasure to especially mention my wife Gesine Arndt and her understanding and patience with me. Without her great moral support it would have been a much harder time. I also owe gratitude to my friend Stephen Reinfranck from SMR Doublebasses who hosted me several times during business trips to the United States.

Contents

List of Figures

List of Tables

Nomenclature

Calligraphic Letters

$\mathcal{A}$ . . . . accuracy, percentage

$\mathcal{E}$ . . . . error

Greek Letters .

α_c . . . . deflection angle of the con rod with respect to the horizontal (crank drive), radians

α_s . . . . surface deflection angle, degree

β . . . . coefficient of thermal expansion, K^{-1}

χ . . . . molar fraction, $\mathrm{mol\,mol}^{-1}$

χ_v . . . . molar fraction of the vapor, $\mathrm{mol\,mol}^{-1}$

ΔH . . . . fill level decrease due to evaporation, m

Δh_v . . . . latent heat of vaporization, $\mathrm{J\,kg}^{-1}$

Δp_{mag} . . . . magnitude of the pressure drop, Pa

δ_0 . . . . initial thermal boundary layer thickness, m

δ_s . . . . STOKES boundary layer thickness, m

δ_t . . . . thermal boundary layer thickness, m

$\eta_{mn} = \omega/\omega_{mn}$ . . . frequency ratio

Γ . . . . Gamma function

γ . . . . damping ratio

Γ_0 . . . . stationary free surface

Γ_S . . . . moving free surface

κ . . . . surface curvature, m^{-1}

κ_c . . . . drive ratio (crank drive)

Λ . . . . logarithmic decrement

Λ_b . . . . logarithmic decrement at the bottom

Λ_s . . . . logarithmic decrement at the side walls

Λ_t . . . . logarithmic decrement at the surface

μ . . . . dynamic viscosity, $\mathrm{Pa\,s}$

ν . . . . kinematic viscosity, $\mathrm{m}^2\,\mathrm{s}^{-1}$

ω	angular velocity, $\mathrm{s^{-1}}$
ω_c	angular velocity of the driving side (crank drive), $\mathrm{s^{-1}}$
ω_{mn}	angular velocity based on the natural frequency, $\mathrm{s^{-1}}$
Φ	velocity potential, $\mathrm{m^2\,s^{-1}}$
ϕ_s	phase shift, radians
ψ_{GHe}	molar fraction of helium dissolved in liquid, $\mathrm{mol\,mol^{-1}}$
σ	surface tension, $\mathrm{N\,m^{-1}}$
τ	characteristic sloshing time scale, s
τ_{ij}	viscous stress tensor, Pa
Θ	characteristic temperature difference, K
ε_{mn}	roots of the BESSEL function J_m
φ_c	angle of rotation (crank drive), radians
φ_d	slope (damping), radians
ϖ_{sub}	degree of subcooling
ϱ	density, $\mathrm{kg\,m^{-3}}$
ϱ_{ref}	reference density, $\mathrm{kg\,m^{-3}}$
ϑ	temperature, K
ϑ_0	initial temperature at the beginning of sloshing, K
ϑ_{amb}	ambient temperature, K
ϑ_{gas}	temperature of the pressurant gas, K
ϑ_{lid}	inner wall temperature at the tank lid, K
ϑ_{ref}	reference temperature, K
ϑ_{sat}	saturation temperature, K
$\vartheta_{\mathrm{sat,0}}$	saturation temperature for $p = p_0$ ($t = 0$), K
ζ	displacement of the free liquid surface, m

Roman Letters

A	contour of the fluid element, $\mathrm{m^2}$
A_Γ	surface area of the free liquid surface, $\mathrm{m^2}$
a_y	lateral acceleration, $\mathrm{m\,s^{-2}}$
a_z	axial acceleration, $\mathrm{m\,s^{-2}}$
A_{mn}	arbitrary constant
B	wave amplitude ratio
B_{mn}	arbitrary constant
C	fitting constant
c	compression rate, $\mathrm{s^{-1}}$
c_p	specific heat capacity at constant pressure, $\mathrm{J\,kg^{-1}\,K^{-1}}$
c_v	specific heat capacity at constant volume, $\mathrm{J\,kg^{-1}\,K^{-1}}$

D'_e . . . . diffusion coefficient, $\mathrm{m^2\,s^{-1}}$

D_e . . . . effective diffusivity, $\mathrm{m^2\,s^{-1}}$

d_s . . . . damping rate, $\mathrm{m\,s^{-1}}$

D_T . . . . thermal diffusivity, $\mathrm{m^2\,s^{-1}}$

D_v . . . . vapor diffusion rate, $\mathrm{m^2\,s^{-1}}$

dm . . . . mass of fluid element, kg

dp_{condens} . . . . . . pressure difference provoked by condensation, Pa

$dp_{\mathrm{gas\,temp}}$ . . . . . pressure difference provoked by ullage cooling down, Pa

$dp_{\mathrm{press\,drop}}$ . . . . . pressure difference provoked by the pressure drop, Pa

dE . . . . change of energy, J

F . . . . volume force, N

f . . . . excitation frequency, Hz

F_0 . . . . initial force, N

f_j . . . . specific body force, $\mathrm{N\,kg^{-1}}$

F_{res} . . . . resulting force, N

g . . . . acceleration due to gravity on Earth, $\mathrm{m\,s^{-2}}$

H . . . . fill level, m

h . . . . height of the ullage region, m

H_e . . . . enthalpy, J

h_e . . . . specific enthalpy, $\mathrm{J\,kg^{-1}}$

$h_{e,\mathrm{condens}}$ . . . . . specific enthalpy of the condensing vapor, $\mathrm{J\,kg^{-1}}$

$h_{e,\mathrm{evap}}$ . . . . specific enthalpy of the evaporating liquid, $\mathrm{J\,kg^{-1}}$

$h_{e,\mathrm{in}}$ . . . . specific enthalpy of the tank entering mass, $\mathrm{J\,kg^{-1}}$

$h_{e,\mathrm{out}}$ . . . . specific enthalpy of the tank leaving mass, $\mathrm{J\,kg^{-1}}$

I_s . . . . specific momentum, s

J . . . . characteristic mass flux, $\mathrm{kg\,m^{-2}\,s^{-1}}$

j . . . . vaporization rate, $\mathrm{kg\,m^{-2}\,s^{-1}}$

J_m . . . . Bessel function of the first order

K . . . . damping coefficient

k . . . . thermal conductivity, $\mathrm{W\,m^{-1}\,K^{-1}}$

K_{deep} . . . . damping coefficient for deep tanks

K_{dome} . . . . dome factor

k_H . . . . HENRY constant, Pa

k_s . . . . spring constant, $\mathrm{N\,m^{-1}}$

L . . . . characteristic length, m

L_c . . . . con rod length (crank drive), mm

L_T . . . . tank height, m

$\dot{m}$ . . . . mass flow rate, $\mathrm{g\,s^{-1}}$

$\dot{m}_{\mathrm{in}}$	. . . .	tank entering mass flow rate, $\mathrm{kg\,s^{-1}}$
$\dot{m}_{\mathrm{out}}$	. . . .	tank leaving mass flow rate, $\mathrm{kg\,s^{-1}}$
$\dot{m}_p$	. . . .	mass flow rate of the consumed rocket propellant, $\mathrm{kg\,s^{-1}}$
M	. . . .	molar mass, $\mathrm{g\,mol^{-1}}$
m	. . . .	mass, kg
m	. . . .	mode number
m_0	. . . .	initial total mass, kg
m_1	. . . .	final total mass, kg
m_{condens}	. . . . . .	condensation mass, kg
m_{evap}	. . . .	evaporation mass, kg
m_{in}	. . . .	mass entering the system, kg
m_{out}	. . . .	mass leaving the system, kg
m_{press}	. . . .	pressurant mass, kg
m_r	. . . .	rocket mass, kg
m_s	. . . .	sloshing mass, kg
M_v	. . . .	molar mass of the vapor, $\mathrm{g\,mol^{-1}}$
P	. . . .	linear momentum, $\mathrm{N\,s}$
n	. . . .	amount of substance, mol
n	. . . .	wave number
p	. . . .	tank pressure, Pa
p_0	. . . .	initial pressure, Pa
p_{amb}	. . . .	ambient pressure, Pa
p_{crit}	. . . .	critical pressure, Pa
p_{GHe}	. . . .	helium partial pressure, Pa
$p_{\mathrm{GN_2}}$	. . . .	partial pressure of nitrogen vapor, Pa
p_{max}	. . . .	maximum pressure, Pa
p_{min}	. . . .	minimum pressure, Pa
p_{ref}	. . . .	reference pressure, Pa
p_{sat}	. . . .	saturation pressure, Pa
p_v	. . . .	vapor partial pressure, Pa
$\dot{Q}$	. . . .	heat flux, W
$\dot{q}$	. . . .	specific heat flux, $\mathrm{W\,m^{-2}}$
$\dot{Q}_{\mathrm{heat}}$	. . . .	heat flux flowing into the tank, W
Q	. . . .	heat (energy), J
Q_{tot}	. . . .	total heat (energy), J
R	. . . .	tank radius, m
R^2	. . . .	coefficient of determineation
R_s	. . . .	specific gas constant, $\mathrm{J\,kg^{-1}\,K^{-1}}$

t	time, s
U	characteristic velocity, $\mathrm{m\,s^{-1}}$
u_Γ	velocity of the phase interphase, $\mathrm{m\,s^{-1}}$
u_in	velocity of the tank entering mass, $\mathrm{m\,s^{-1}}$
u_out	velocity of the tank leaving mass, $\mathrm{m\,s^{-1}}$
U_e	internal energy, J
u_e	specific internal energy, $\mathrm{J\,kg^{-1}}$
u_i	i^th component of the velocity vector $\mathbf{u}$, $\mathrm{m\,s^{-1}}$
u_v	interfacial vapor velocity, $\mathrm{m\,s^{-1}}$
u_{ex}	effective exhaust velocity, $\mathrm{m\,s^{-1}}$
u_r	rocket velocity, $\mathrm{m\,s^{-1}}$
$\dot{V}$	volume flow rate, $\mathrm{m^3\,s^{-1}}$
V	volume, $\mathrm{m^3}$
W	mechanical work, J
x	mass fraction, $\mathrm{kg\,kg^{-1}}$
x_v	mass fraction of the vapor
y_A	excitation amplitude, m
y_h	homogeneous solution, m
y_p	particular solution, m

Subscript

condens	condensation
evap	evaporation
flush	flushing phase
in	entering the control volume
out	leaving the control volume
press	pressurization phase
sat	saturation
slosh	sloshing phase
tot	total
f	after flushing
$f+p$	after flushing and pressurization
i	component: 1 = radial, 2 = circumferential, 3 = axial (z)
k	number of layer in the liquid and in the ullage
L	liquid phase
p	pressurization
U	ullage phase
GHe	GHe gas phase
GN2	GN_2 gas phase

Superscript

$'$	during sloshing
(I), (II)	beginning, end
$*$	dimensionless quantity

Vectors

$\Delta\mathbf{u}_r$	change in rocket velocity, $\mathrm{m\,s^{-1}}$
$\mathbf{A}$	vector potential function, $\mathrm{m\,s^{-1}}$
$\mathbf{F}_t$	thrust, N
$\mathbf{g}$	acceleration vector due to gravity, $\mathrm{m\,s^{-2}}$
$\mathbf{I}$	unit tensor
$\mathbf{I}_{\mathrm{tot}}$	total momentum, N s
$\mathbf{I}_r$	momentum of the rocket, N s
$\mathbf{I}_t$	momentum emerging from the thrust, N s
$\mathbf{n}$	normal vector
$\mathbf{T}$	stress tensor, Pa
$\mathbf{u}$	velocity vector, $\mathrm{m\,s^{-1}}$

Coordinates System

θ	coordinate in circumferential direction (cylinder)
r	coordinate in radial direction (cylinder)
z	coordinate in flight direction (cylinder)
x	coordinate perpendicular to the sloshing direction (cartesian)
y	coordinate in sloshing direction (cartesian)
z	coordinate in flight direction (cartesian)

Nondimensional Numbers

$\mathbf{Bo} = \frac{\varrho g L^2}{\sigma}$	BOND number
$\mathbf{Ev} = \frac{\Delta h_v \, J \, L}{k\,\Theta}$	evaporation number
$\mathbf{Fr} = \frac{\mathbf{U^2}}{\mathbf{g\,L}}$	FROUDE number
$\mathbf{Ga} = \frac{g\,L^3}{\nu^2}$	GALILEI number
$\mathbf{Gr} = \frac{g\,\beta\,\Theta\,L^3}{\nu^2}$	GRASHOF number
$\mathbf{Ja} = \frac{c_p\,\Theta}{\Delta h_v}$	JACOBS number
$\mathbf{Pr} = \frac{\mu\,c_p}{k}$	PRANDTL number
$\mathbf{Ra} = \frac{g\,\varrho_{\mathrm{ref}}\,\beta\,\Theta\,L^3\,c_p}{k\,\mu 0}$	RAYLEIGH number
$\mathbf{Re} = \frac{\varrho_{\mathrm{ref}}\,U\,L}{\mu}$	REYNOLDS number
$\mathbf{R} = \frac{\varrho_{\mathrm{ref}}}{\varrho_U}$	density ratio
$\mathbf{Sc} = \frac{\mu}{\varrho_{\mathrm{ref}}\,D_v}$	SCHMIDT number

$\mathbf{St} = \frac{L}{\tau U}$ . . . . .   STROUHAL number

$\mathbf{We} = \frac{\varrho_{ref} U^2 L}{\sigma}$ . .   WEBER number

Abbreviations   .

CFD	computational fluid dynamics
ESC-A	Étage Supérieur Cryotechnique A
ESC-B	Étage Supérieur Cryotechnique B
GH_2	gaseous hydrogen
GHe	gaseous helium
GN_2	gaseous nitrogen
GOX	gaseous oxygen
ISS	international space station
LH_2	liquid hydrogen
LN_2	liquid nitrogen
LOX	liquid oxygen
POM	polyoxymethylene
ZARM	Center of applied space technology and microgravity

Chapter 1

INTRODUCTION

This work is dedicated to a - in general mostly unwanted - dynamic fluid behavior that is typically known as liquid sloshing. There exist many applications in civil and industrial engineering, in which this effect is of major interest. Frequently, liquid sloshing has been studied by designers of road and ship tankers as well as by geologists and architects to study its impact. Examples include large dam walls affected by underground motion and large liquid containers mounted on top of multistory buildings used as earthquake protection systems. Furthermore, the field of applications is enhanced by the aerospace industry where the understanding of liquid sloshing is of fundamental importance to design appropriate propellant tanks for spacecrafts. Liquid sloshing is defined as more or less periodical motion of the free liquid surface inside a closed reservoir[1]. It is provoked by any disturbances such as vibrations, acceleration changes or rolling and pitching movements of the tank. Depending on the excitation, this can cause different types of liquid motion including symmetric, asymmetric, swirling and chaotic sloshing modes.

This chapter starts with a compact review of the past. Afterwards, the application of cryogenic propellants including the risks compared to the utilization of storable liquids is discussed. The central part of this chapter deals with the principle question that shall be answered by this work. Furthermore, the state of the art related to cryogenic liquid sloshing is presented before the chapter closes with the motivation and the objectives for this work.

For centuries, one dream of Humankind is the exploration of the space - leaving Earth's atmosphere and beyond. Long before aeronautics and astronautics could be established as technical applications, these partially dreamy issues were primarily discussed by philosophers and litterateurs who cast their utopistic ideas into techno-romantic tales trying to influence the modern age world view. Well known ambassadors of this movement are LEONARDO DA VINCI, GIORDANO BRUNO, JOHANNES KEPLER, and JULES VERNE, to name only a few of them. An example is provided in figure 1.1 showing the attempt to reach the moon by using a large gun bullet.

---

[1]Sloshing affects the liquid of a certain distance below the surface as well.

Figure 1.1: The shot to the Moon taken from the novel *From the Earth to the Moon* by Jules Verne written in 1865. The illustrations are taken from [56].

Driven by the cold war, space flights became of major interest during the 1950's and 1960's, when American and Soviet space programs were strongly supported by their governments, focusing on the manned exploration of Earth's orbit and the Moon. Highlights in this age were the first artificial earth satellite SPUTNIK launched by the USSR, the first manned space flight by YURI GAGARIN, and the American APOLLO program to perform the first manned landing on the Moon. Later, in the 1980's and 1990's, the reusable Space Shuttle and the International Space Station (ISS) shaped the face of modern astronautics. For unmanned missions, the European space launcher Ariane 5 advanced as one of the most successful carrier systems for satellite shipment into Earth's orbit.

Beating Earth's gravity field to reach the orbital position defines the main task of any launcher system that has been employed or will be designed in future. According to NEWTONs third law of motion, a rocket works on the principle providing a continuous ejection of hot gases in one direction to cause a thrust $\mathbf{F}_t$ and therefore a steady acceleration in the opposite direction. Thus, the momentum balance of the rocket must satisfy

$$\frac{d\,\mathbf{I}_{\text{tot}}}{dt} = \frac{d\,\mathbf{I}_r}{dt} + \frac{d\,\mathbf{I}_t}{dt} \tag{1.1}$$

where $\mathbf{I}_{\text{tot}}$ gives the total momentum. On the one side, the momentum of the rocket is defined as

$$\mathbf{I}_r\left(t\right) = m_r\left(t\right)\,\Delta\mathbf{u}_r\left(t\right)\,. \tag{1.2}$$

Here, m_r is the actual rocket mass including the varying propellant mass and $\Delta\mathbf{u}_r$ is the change in rocket velocity. With the effective exhaust velocity u_{ex}, equation (1.2) gives in scalar notation

Table 1.1: The specific momentum I_s of actual and outdated launcher systems [18, 44].

launcher	stage/engine	propellant	kind	I_s [s]
Saturn V	F-1	LOX/RP-1	semi-cryogenic	265
	J-2	LOX/LH$_2$	cryogenic	424
Ariane 4 (AR44 LP)	2	UH25/N$_2$O$_4$	storable	294
	3	LOX/LH$_2$	cryogenic	444
Ariane 5	EPS	N$_2$O$_4$/MMH	storable	231
	ESC-A	LOX/LH$_2$	cryogenic	447
Space Shuttle	SRB	Al/NH$_4$ClO$_4$	solid	295
	SSME	LOX/LH$_2$	cryogenic	455

$$-u_{ex}\frac{dm_r}{dt} = m_r\frac{du_r}{dt} \tag{1.3}$$

leading to the well known ZIOLKOWKY rocket equation

$$\Delta u_{12} = u_{ex}\ln\left[\frac{m_0}{m_1}\right] \tag{1.4}$$

where the initial total mass is denoted as m_0 and the final total mass is defined as m_1.

On the other side, the momentum emerging from the engine thrust $\mathbf{F}_t = v_e\,\dot{m}_p$ is defined as

$$\mathbf{I}_t\left(t\right) = \int_0^t \mathbf{F}_t\,dt. \tag{1.5}$$

According to equation (1.1), an payload extension requires a more powerful thrust provided by the engine. However, the specific momentum is a convenient measure to gauge the efficiency of rocket propellants. The specific momentum is defined as propellant weight related thrust, so that

$$I_s = \frac{|\mathbf{I}_t|}{a_z\int_0^t \dot{m}_p\,dt} \tag{1.6}$$

where $a_z = g$ is the gravitational acceleration on Earth and $\dot{m}_p$ is the actual propellant mass flow rate of the rocket. Although the specific momentum is a characteristic of the propellant system, its exact value will vary to some extent with the operating conditions and design of the rocket engine. The higher the specific momentum, the more energy is stored in the propellant and therefore the more payload can be carried by the launcher. For constant thrust and constant propellant flow, equation (1.6) simplifies yielding

$$I_s = \frac{|\mathbf{F}_t|}{g\,\dot{m}_p}. \tag{1.7}$$

(A) Ariane 5   (B) cryogenic upper stage

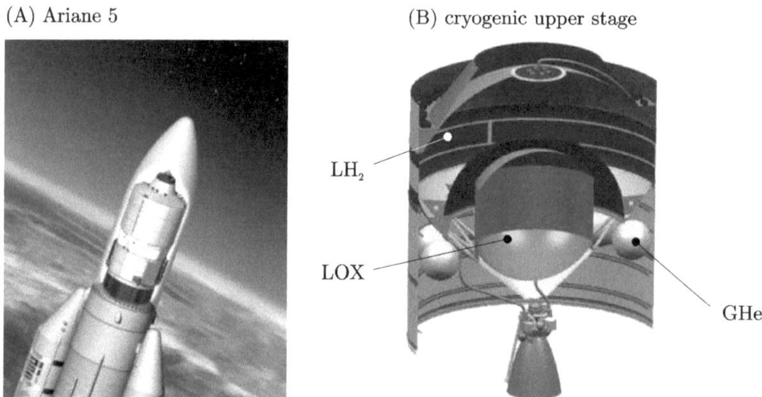

Figure 1.2: (A) The European space launcher Ariane 5 (illustration is provided by ESA). (B) Tank configuration of the cryogenic upper stage ESC-A [17] including the reservoirs for liquid hydrogen (LH$_2$), liquid oxygen (LOX) and gaseous helium (GHe).

For the general consideration of rocket propulsion systems during the flight where start and stop transients can be neglected, equation (1.7) is most appropriate.

Table 1.1 provides information about the specific momentum of actual and outdated launcher systems including Saturn V, Ariane 4, Ariane 5 and the Space Shuttle. In terms of I_s, obviously cryogenic systems provide the higher efficiency in comparison to storable propulsion systems driven typically by hydrazine[2] based propellants. For this reason, cryogenic engines are preferentially applied in rocket upper stages that are designed to be launched in lower orbits. However, a major disadvantage of cryogenic propellants is their notably low saturation temperature producing high amounts of boil-off. This represents a big challenge in terms of long time storage. The cryogenic upper stage ESC-A of the European space launcher Ariane 5 is shown in figure 1.2. Beside the tanks to store liquid oxygen (LOX) and helium (He), the axisymmetric reservoir for liquid hydrogen (LH$_2$) is the largest tank compartment of this stage. Initially, this tank is pressurized up to a certain system pressure that is in the order of 300 kPa, while the tank is approximately 95% full of cryogenic propellant.

During the ascent phase when the rocket executes several flight maneuvers such as rolling and pitching in order to reach Earth's orbit, liquid propellants tend to more or less periodic surface movements within the tank. Commonly, this is called liquid sloshing, which is typically caused by variations of the linear and angular acceleration in x, y and z direction. For most applications concerning propellant storage in rocket tanks, liquid sloshing is an undesired side effect causing unwanted mechanical and thermodynamical reactions in the rockets feed system. In this connection, two aspects are of major importance: damping and sudden pressure drops.

---

[2]Hydrazine is a storable nitrogen based propellant that is often utilized with nitric acid as oxidizer.

Damping

Other than using solid or gaseous propellants, the utilization of liquid propellant imply the permanent occurrence of liquid motion that can not be completely prevented in rocket tanks. A notable property of liquid sloshing is the ability to dissipate energy as a result of viscous stresses within the liquid and between the liquid and the wall boundary. Commonly, this property is called damping. Among others, the damping characteristic of a liquid significantly influences the sloshing dynamics and in particular the maximum wave amplitude. In general, the higher the damping, the less dominant are the sloshing dynamics. Damping is directly connected to the liquid viscosity ν, to the tank size R, to the acceleration level a_z and for some extend to the tank fill level H. Since the damping level for a given tank/propellant combination is fixed during the ascent phase[3], the sloshing dynamics can be reduced by according tank design features such as the application of horizontal ring baffles. Ring baffles represent a substantial obstacle for the liquid motion during sloshing. However, the knowledge about the damping characteristics may help to improve the design of suitable tank solutions particularly for future upper stages.

Pressure drops

By the utilization of cryogenic propellants, another important aspect is given by the impact of sloshing on the coupling between the heat and mass transfer at the phase interface between liquid and vapor. Background of this assumption is the fact that previously a sudden characteristic pressure drop occurred in the hydrogen tank of the Ariane 4 upper stage due to propellant sloshing during the ascent phase. As a matter of fact, a pressure drop might be critical compromising the structural stability of the tank. As mentioned before, the tank pressure is approximately $p = 300$ kPa. Due to certain flight maneuvers, the tank content is initiated to slosh. Within seconds, the pressure in the tank drops by 5 to 10%. It is assumed that the characteristic pressure drop phenomenon is caused by condensation effects that are provoked by the mixing of liquid at the free surface with liquid from the bulk.

Both aspects, the damping characteristics as well as the thermodynamic phenomena - such as the characteristic pressure drop - are of major importance since these effects may influence the tank design of future cryogenic upper stages. Furthermore, this might also be of interest by considering micro-gravity conditions particularly for engine shut-down and restart maneuvers in orbit. Here, propellant sloshing preferably may be considered as settling effect where the vector of the motion is different, but the occurring fluid dynamics as well as thermodynamic effects may appear in a similar manner.

---

[3]The tank is assumed to be filled to more than half of its volume with propellant.

1.1   State of the Art

The theoretical consideration of sloshing liquids in closed containers can be traced back to LAMB [40] who presented the mathematical description of oscillating surface waves in cylindrical tanks based on the potential theory using BESSEL functions. Based on his approach, he could identify different sloshing modes showing a strong dependency on the excitation frequency. Viscous effects in cylindrical tanks with flat bottoms were considered firstly by MILES [46] as well as CASE & PARKINSON [22]. They assumed that viscous energy dissipation can be referred to a small boundary layer in the vicinity of the tank walls. In general, the dominant parameters that influence viscous damping effects in liquids can be characterized by the GALILEI number defined as the ratio of gravitational forces to viscous forces so that $\mathbf{Ga} = a_z R^3/\nu^2$. However, an experimental proof was performed by STEPHENS $et\ al.$ [55] conducting sloshing tests in a cylindrical tank using water. MIKISHEV & DOROZHKIN [45] studied firstly spherical bottom geometries. Particularly for fill levels in the order of the tank size and below, they could identify a significant influence of the tank geometry on the damping characteristics of the liquid. Referring to the American aerospace agency, the most complete compendium in this period concerning liquid sloshing in rocket propellant tanks were composed by BAUER [15] as well as by ABRAMSON [1] for NASA. Particularly the latter work contains approaches for analogous models to describe liquid sloshing. This includes spring mass and pendulum models where the liquid content is substituted by an oscillating mass point. Viscous properties were added by appropriate dashpot elements, so that the models are reduced to ordinary differential equations to describe the complex liquid system. Recently, this work was rewritten by DODGE [27] and later by IBRAHIM [37], in which numerical methods for the analogous models were updated using more efficient simulation codes to compute the occurring sloshing forces and the natural frequencies particularly under the consideration of complex tank geometries.

With regard to the planned cryogenic upper stage for the European space launcher Ariane 5, experimental sloshing tests with storable liquids were conducted by ROYON-LEBEAUD $et\ al.$ [51, 52]. They studied large amplitude sloshing and swirling waves in square-base and cylindrical tanks. ARNDT & DREYER [3, 6] performed sloshing experiments to test the influence of the bottom geometry on the damping characteristics of the liquid. They considered flat, spherical and convex bottom geometries using water to verify the damping models for storable liquids developed by MIKISHEV & DOROZHKIN [45] and STEPHENS $et\ al.$ [55]. The numerical results showed quite good agreement except for a flat bottom geometry where the numerical damping results performed with the commercial CFD software FLOW3D[4] over-predict the experimental results by approximately a factor of 2.

In terms of cryogenic propellants, fundamental research was done on behalf of NASA in the 1960's and 1970's. One of the first cryogenic rocket engines that was developed by ROCKET-

---

[4]FLOW3D Version 9.1.1. by FLOW SCIENCE INC.

DYNE is the J-2. Amongst others, this engine was applied on the third stage of the Saturn V rocket, the motor of the Apollo space program. The AS-203 experiment [57] that was performed on the V/S-IVB stage of a Saturn-I-B rocket, was conducted to verify the cryogenic propellant control system in consideration of the application on the Saturn V upper stage. Equipped with a variety of sensors, the main focus of this mission was to study cryogenic propellant characteristics during the ascent phase and in a low gravity environment.

To enhance the understanding of the thermodynamics within the framework of cryogenic propulsion, much work was done on fields of pressurization [10, 24, 50] and stratification [13, 14, 24, 42]. Recent numerical investigations in this area are conducted by GRAYSON et al. [31, 32], HIMENO et al. [35, 34] and LACAPERE et al. [39]. They studied the phase change at the free liquid surface in cryogenic liquids under the impact of sloshing.

MORAN et al. [48] performed full size ground sloshing experiments in pressurized spherical tanks using liquid hydrogen LH_2. Varying the excitation frequency, amplitude and ullage volume, the thermodynamic response under the impact of Earth's gravity could be characterized. It was found that particularly in the vicinity of the first natural frequency, sloshing effects have a major influence on the tank pressure development that may lead into ullage collapse. Furthermore, they could show that the pressure drop can be prevented by using helium as non-condensable pressurant gas. The pressure drop phenomenon on laboratory size scale was recently observed by LACAPERE et al. [39] studying laterally excited liquid nitrogen LN_2 in a $R = 0.095$ m cylindrical tank. The pressure drop could be numerically determined with a customized version of the commercial CFD software FLUENT. For the first asymmetric and symmetric sloshing modes, DAS & HOPFINGER [26] observed the characteristic pressure drop by using the engineering fluids FC-72 and HFE7000. This coheres with studies performed by HOPFINGER & DAS [36] who used FC-72 and HFE7000 as well. They could express the observed pressure variation in terms of an effective diffusion coefficient, the JACOB number and the temperature gradient in the boundary layer in the vicinity of the free surface. Experimental studies under variation of the pressurization are conducted by ARNDT et al. [7, 8] showing the influence of a non-condensable pressurant gas on the pressure drop phenomenon.

1.2   Motivation and Objectives

The motivation of this work is coupled to the development of a new cryogenic upper stage for the European space launcher Ariane 5. Particularly the layout of the propellant tank compartment to store liquid hydrogen (LH_2) may be influenced by the results that are discussed in this work. It is of fundamental importance to enhance the understanding of the fluid-dynamics of the cryogenic propellant and the thermodynamical effects that might occur under the impact of sloshing within the tank.

The understanding of the viscous damping for storable liquids such as water or engineering fluids is a widely explored field [6, 45, 52, 55]. In the current work, the damping model based on the GALILEI number **Ga** will be tested for cryogenics and in particular for liquid nitrogen LN_2 as substitute for liquid hydrogen[5] LH_2. For a constant tank size, the fill level will be varied to emphasize the influence of the bottom geometry on the damping characteristics in an axis-symmetric tank with a spherical bottom shape.

Only few experimental investigations can be found in the literature concerning cryogenic propellant sloshing in gravity dominated environments that would reflect the conditions during the ascent phase [26, 39, 48]. Particularly the consideration of variable initial conditions represents a field of further investigation that may be important for the design of future upper stages.

The objectives of this work include the installation of an appropriate laboratory scale test facility in order to perform cryogenic sloshing tests with liquid nitrogen. Under the requirements of different initial conditions, the impact of the sloshing liquid shall be tested to study the changes of the corresponding thermodynamical conditions in the tank to identify the characteristic pressure drop phenomenon. For this purpose, the tank shall be pressurized by three different methods including

- *self-pressurization* where the tank is pressurized by means of the parasitic heat flow that continuously leaks into the system,

- *external nitrogen pressurization* where gaseous nitrogen from an external gas reservoir is fed into the tank, and

- *external helium pressurization* where gaseous helium from an external gas reservoir is fed into the tank.

The helium experiments are of particular interest, since helium is a non-condensable gas in a nitrogen environment. It is expected to reduce the pressure drop under the impact of sloshing. The obtained data is presented in nondimensional form in order to allow the up-scaling to other dimensions including previous experimental results as well as to enable predictions for the full size application. Previous data by MORAN *et al.* [48], LACAPERE [39] and DAS & HOPFINGER [26] performing experiments with higher initial pressures and other liquids are considered to confirm the actual results.

This work shall help to enhance the understanding of handling cryogenic liquids particularly when they are stored in upper stage rocket tanks. Furthermore, this work shows the limits of simple laboratory size experiments in order to simulate the complex full size geometry of upper stage tanks. Derived from actual results, it also suggest some motivation for future activities on this field.

---

[5]In the current case, the security guidelines for laboratory purpose of the University of Bremen prohibit the utilization of liquid hydrogen for quantities of approx. 20 liter without extensive modifications of the infrastructure.

Chapter 2

THEORETICAL BACKGROUND

This chapter is dedicated to the theoretical background to describe the occurring effects related to liquid sloshing. It starts with the introduction of the governing transport equations in accordance to explain the fluid dynamical problem. From these equations the corresponding characteristic numbers are derived to apply an adequate scaling. Furthermore, the fluid motion in the container is described by means of the potential theory. This theory is enhanced by a vortex potential function in order to consider the viscous property of the fluid as well [46, 45, 55]. The pressure drop phenomenon is theoretically described by an approach of DAS & HOPFINGER [26] by considering an effective diffusion coefficient within the liquid.

Figure 2.1 shows a typical cylindrical propellant tank with a spherical bottom shape as often found in actual and previous applications [1, 17, 27] with a liquid fraction (L) and an ullage (U).

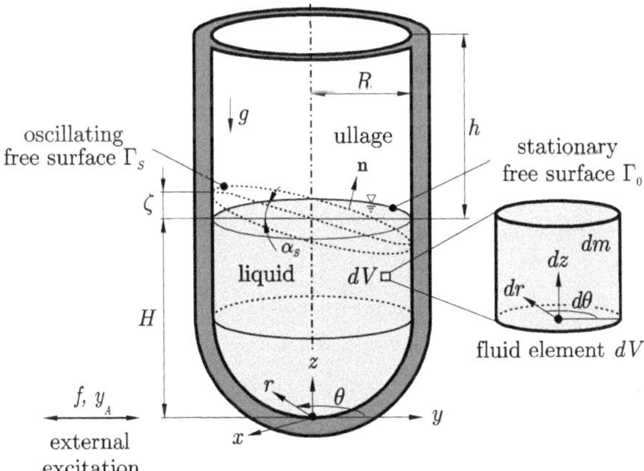

Figure 2.1: Schematic illustration of the tank including the definition of a single fluid element dV of mass dm. The tank consists of a liquid and an ullage phase. The axes are defined in cylindrical and cartesian coordinates, while the origin is set to the tank bottom.

The tank is excited in y direction, whereas f is the frequency and y_A is the amplitude of the oscillation. The z axis points contrary to the gravity vector $\mathbf{g} = (0, 0, -g)^T$, while the x direction points perpendicular to the excitation direction. Internally in terms of cylindrical coordinates, the radial direction is defined as r, the circumferential direction is θ, and again, z points from bottom to the top along the symmetry axis. The origin of the cylindrical coordinate system is set to the tank bottom. The stationary free surface is denoted as Γ_0, while the moving free surface is denoted as Γ_S. The normal vector perpendicular to the free surface is defined as $\mathbf{n}$.

The cylindrical tank has a radius of R and is filled with liquid propellant up to a certain fill level H. Due to the excitation, the propellant content is forced into a sloshing motion, whereas the displacement of the oscillating liquid surface can be approximated by $\zeta = R \tan \alpha_s$ with respect to the stationary free surface. Here, α_s is the deflection angle of the free surface. The height of the ullage region is defined as h, so that the entire height of the tank is $L_T = H + h$. As described at the end of this chapter, the liquid motion can appear in different sloshing modes, which strongly depend on the excitation, the tank size, and the liquid properties.

2.1 Governing Equations

Typical fluid dynamical problems can be described by the three transport equations that are given by the conservation of mass, the conservation of momentum and the conservation of energy. In the following, these equations are introduced in general based on [19, 12, 23, 40, 49]. The local velocity of the fluid element dV (see figure 2.1) with the mass dm is defined as $\mathbf{u} = (u_r, u_\theta, u_z)^T$ in cylindrical coordinates. In order to satisfy the conventional definition, the liquid is considered as continuum with a free surface. In the following, the governing equations are introduced describing a one-phase system with liquid.

2.1.1 Conservation of Mass

The basic principle of the mass conservation includes that the mass dm of the fluid element dV with the density ϱ is assumed to be constant. Therefore, the time variation of the mass equals the difference between inlet and outlet fluxes - or in general - the mass flow rate $\dot{m}$, so that

$$\frac{dm}{dt} = \dot{m} \, . \tag{2.1}$$

The time variation of the mass in the considered fluid element dV is defined as

$$\frac{dm}{dt} = \frac{\partial}{\partial t} \int_V \varrho \, dV = \int_V \frac{\partial \varrho}{\partial t} \, dV \, . \tag{2.2}$$

The net mass flux along the contour dA of the fluid element dV is defined as

$$\dot{m} = - \oint_A \varrho\, u_i\, dA_i \tag{2.3}$$

where u_i gives the ith velocity component of the velocity vector $\mathbf{u}$ at the fluid element contour A_i. Here, $i = 1$ corresponds to the radial direction r, $i = 2$ corresponds to the circumferential direction θ, and $i = 3$ corresponds to the axial direction z. However, setting equation (2.2) and (2.3) into equation (2.1) gives the integral form of the mass balance

$$\int_V \frac{\partial \varrho}{\partial t}\, dV + \oint_A \varrho\, u_i\, dA_i = 0\,. \tag{2.4}$$

Applying GAUSS theorem for volume calculation, the contour integral in equation (2.4) can be transformed into the volume integral

$$\int_V \frac{\partial \varrho}{\partial t}\, dV + \int_V \frac{\partial \varrho\, u_i}{\partial x_i}\, dV = 0\,. \tag{2.5}$$

For infinitesimal volumes $V \to 0$, the differential form of the mass balance yields

$$\frac{\partial \varrho}{\partial t} + \nabla \cdot (\varrho\, \mathbf{u}) = 0\,. \tag{2.6}$$

2.1.2   Conservation of Momentum

The conservation of momentum is derived from NEWTON's second law explaining that the time rate of the change of the linear momentum is proportional to the net forces F_j acting on the fluid element dV with the mass dm yielding

$$\frac{d\,(m\, u_j)}{dt} = \sum F_j \tag{2.7}$$

where the linear momentum P is defined as the product of the fluid mass m and the local fluid velocity $\mathbf{u}$, so that $P \equiv m\,\mathbf{u}$. Thus, the left hand side of equation (2.7) reads

$$\frac{d\,(m\, u_j)}{dt} = \int_V \frac{\partial\,(\varrho\, u_j)}{\partial t}\, dV + \oint_A \varrho\, u_i\, u_j\, dA_i \tag{2.8}$$

consisting of a time dependent unsteady term and a term including the convective momentum transport. On the right hand side of equation (2.7), the net forces give

$$\sum F_j = - \underbrace{\oint_A p\, dA_j}_{(I)} + \underbrace{\oint_A \tau_{ij}\, dA_i}_{(II)} + \underbrace{\int_V \varrho\, f_j\, dV}_{(III)}\,, \tag{2.9}$$

where (I) gives the static pressure gradient, (II) gives the viscous momentum transport, and (III) gives the specific body forces f_j acting on the fluid element. In consequence, the momentum balance in its integral form reads

$$\int_V \frac{\partial\left(\varrho\, u_j\right)}{\partial t}\, dV + \oint_A \varrho\, u_i\, u_j\, dA_i = - \oint_A p\, dA_j + \oint_A \tau_{ij}\, dA_i + \int_V \varrho\, f_j\, dV\,. \tag{2.10}$$

Applying the GAUSS theorem for volume calculation, the contour integrals can be converted into volume integrals. Then, equation (2.10) yields

$$\int_V \frac{\partial\left(\rho\, u_j\right)}{\partial t}\, dV + \int_V \frac{\partial\left(\rho\, u_i\, u_j\right)}{\partial x_i}\, dV = - \int_V \frac{\partial p}{\partial x_j}\, dV + \int_V \frac{\partial \tau_{ij}}{\partial x_i}\, dV + \int_V \varrho\, f_j\, dV\,. \tag{2.11}$$

For an incompressible Newtonian fluid, the viscous stress tensor τ_{ij} reduces to

$$\tau_{ij} = \mu \left[\frac{\partial u_j}{\partial x_i} + \frac{\partial u_i}{\partial x_j} \right] \tag{2.12}$$

with μ representing the dynamic viscosity. Then, the differential form of the momentum balance given in equation (2.11) can be rewritten in vector notation

$$\varrho \left[\frac{\partial \mathbf{u}}{\partial t} + (\mathbf{u} \cdot \nabla)\, \mathbf{u} \right] = -\nabla p + \mu\, \nabla^2 \mathbf{u} + \varrho\, \mathbf{g} \tag{2.13}$$

where the specific body forces f_j are replaced by the gravitational acceleration $\mathbf{g} = (0, 0, -g)^T$ representing the buoyancy term.

2.1.3   Conservation of Energy

According to the first law of thermodynamics for closed systems, the fluid element dV can be balanced with the total internal energy $dU_{e,\,\mathrm{tot}}$ stored in the fluid, which is equal to the amount of energy added by heating dQ as well as by mechanical work dW. Hereby, the system can be considered as an isochoric process since the volume of the tank is constant. Thus, the energy balance yields

$$\frac{dQ}{dt} + \frac{dW}{dt} = \frac{dU_{e,\,\mathrm{tot}}}{dt} + \dot{m}_{\mathrm{out}} \left[h_{e\,\mathrm{out}} + \frac{u_{\mathrm{out}}^2}{2} + g\, z \right] - \dot{m}_{\mathrm{in}} \left[h_{e\,\mathrm{in}} + \frac{u_{\mathrm{in}}^2}{2} + g\, z \right]\,. \tag{2.14}$$

For a closed system, the mass flow rates entering the tank $\dot{m}_{\mathrm{in}}$ and leaving the tank $\dot{m}_{\mathrm{out}}$ are zero. The energy term on the left hand side includes the internal energy (I) and the energy flux at the volume contour (II), so that

$$\frac{dU_{e,\,\mathrm{tot}}}{dt} = \underbrace{\int_V \frac{\partial\left(\varrho\, u_e\right)}{\partial t}\, dV}_{(I)} + \underbrace{\oint_A \varrho\, u_e\, u_i\, dA_i}_{(II)} \tag{2.15}$$

where $u_e = U_e/m$ is defined as specific internal energy with respect to the mass. While the internal energy is stored within the fluid, the energy flux at the volume contour corresponds to the mass that may be added or removed from the system. Furthermore, it is more convenient to express the energy stored in the fluid by introducing the enthalpy H_e [11]. The enthalpy is defined as

$$H_e = U_e + pV \tag{2.16}$$

or in terms of the change of enthalpy

$$dH_e = dU_e + pdV + V\,dp\,. \tag{2.17}$$

However, the heat that is added can enter the system only through the volume contour yielding

$$\frac{dQ}{dt} = \oint_A \dot{q}_i\,dA_i \tag{2.18}$$

with $\dot{q} = \dot{Q}/A$ as specific heat flux. Mechanical work is only added indirectly to the system (excitation) and can therefore be neglected. According to EULER's approach, the external excitation does not need to be considered here. Thus,

$$\frac{dW}{dt} = 0\,. \tag{2.19}$$

Setting equation (2.15), (2.18), and (2.19) into equation (2.14) leads to the integral form of the energy balance given by

$$\int_V \frac{\partial(\varrho\,u_e)}{\partial t}\,dV + \oint_A \varrho\,u_e\,u_i\,dA_i = \oint_A \dot{q}_i\,dA_i\,. \tag{2.20}$$

Again, applying the GAUSS theorem for volume calculation, the contour integrals can be converted into volume integrals yielding

$$\int_V \frac{\partial(\varrho\,u_e)}{\partial t}\,dV + \int_V \frac{\partial(\varrho\,u_e\,u_i)}{\partial x_i}\,dV = \int_V \frac{\partial\dot{q}_i}{\partial x_i}\,dV\,. \tag{2.21}$$

The right hand side term can be rewritten by means of FOURIER's law of heat conduction $\dot{q}_i = -k\,\partial\vartheta/\partial x_i$. Regarding that the internal specific energy can be also expressed by the heat capacity for constant volume c_v, so that $u_e = c_v\,\vartheta$, the energy balance for a constant volume reads

$$\frac{\partial(\varrho\,c_v\,\vartheta)}{\partial t} + \frac{\partial(\varrho\,c_v\,\vartheta\,u_i)}{\partial x_i} = k\,\frac{\partial^2\vartheta}{\partial x_i^2} \tag{2.22}$$

where k is the thermal conductivity of the fluid. In vector notation, equation (2.22) can be rewritten

$$\varrho\,c_v\left[\frac{\partial\vartheta}{\partial t} + \mathbf{u}\cdot\nabla\vartheta\right] = k\,\nabla^2\vartheta\,. \tag{2.23}$$

2.2   Scaling Concept for Gravity Dominated Liquid Sloshing

Characteristic numbers represent a convenient method to scale laboratory size experiments to obtain predictions for the full size application [29]. Typical fluid dynamical problems are described by the three transport equations that are introduced in equations (2.6), (2.13), and (2.23). For the considered sloshing case, some assumptions are required to identify the adequate characteristic numbers.

Since the considered fluids in this work are cryogenics, the liquid is assumed to be compressible, so that $c_v \neq c_p$ as shown by the properties in table 3.2. Furthermore, the BOUSSINESQ approximation is applied assuming that effects based on volume expansions caused by temperature variation can be neglected [23], so that $\varrho_L = \varrho_{\text{ref}}$ except in the buoyancy term of the momentum balance where $\varrho = \varrho_{\text{ref}} - \varrho_{\text{ref}}\, \beta\, (\vartheta - \vartheta_{\text{ref}})$. Here, ϱ_{ref} is the reference density corresponding to the liquid density for the reference temperature ϑ_{ref}. Furthermore, β is the coefficient of thermal expansion, a measure for the response to temperature changes in the system. The gas in the ullage satisfies the ideal gas law $p = \varrho_U\, R_s\, \vartheta_U$ where R_s is the specific gas constant of the ullage gas.

To satisfy the rocket tank requirements as two-species system, the liquid and vapor phase are considered separately. While the liquid phase (L) represents a pure species[1] (propellant), the vapor phase in the ullage (U) might be composed of a mixture of the propellant vapor and a non-condensable inert gas that corresponds in most applications to gaseous helium GHe. Thus, the transport equations in the liquid can be rewritten in order to agree with these assumptions, so that

$$\nabla \cdot \mathbf{u}_L = 0\,, \tag{2.24}$$

$$\varrho_{\text{ref}} \left[\frac{\partial \mathbf{u}_L}{\partial t} + (\mathbf{u}_L \cdot \nabla)\, \mathbf{u}_L \right] = -\nabla p + \mu\, \nabla^2 \mathbf{u}_L + \varrho_{\text{ref}}\, \mathbf{g} - \varrho_{\text{ref}}\, \beta\, (\vartheta_L - \vartheta_{\text{ref}})\, \mathbf{g}\,, \tag{2.25}$$

$$\varrho_{\text{ref}}\, c_{p,L} \left[\frac{\partial \vartheta_L}{\partial t} + \mathbf{u}_L \cdot \nabla \vartheta_L \right] = k_L\, \nabla^2 \vartheta_L\,. \tag{2.26}$$

The velocity of the moving phase interphase u_Γ can be expressed by the velocity normal to Γ_S and the vaporization rate j yielding

$$u_\Gamma = \mathbf{u} \cdot \mathbf{n}_\Gamma - \frac{1}{\varrho_{\text{ref}}}\, j\,, \tag{2.27}$$

where $\mathbf{n}_\Gamma$ is the vector normal to the free surface. Furthermore, the vaporization rate (mass flux) gives

$$j = -\frac{1}{\Delta h_v} \left[k_L\, \frac{\partial \vartheta_L}{\partial \mathbf{n}_\Gamma} - k_U\, \frac{\partial \vartheta_U}{\partial \mathbf{n}_\Gamma} \right] \tag{2.28}$$

---

[1] Dissolved gas is neglected here

with the latent heat of vaporization Δh_v. The temperature at the surface corresponds to the saturation temperature ϑ_sat that only depends on the vapor partial pressure p_v, so that

$$\vartheta = \vartheta_\text{sat}\left(p_v\right) . \tag{2.29}$$

The ullage may consist of two species, the vapor and the non-condensable gas GHe. This is considered in the transport equations of the gas phase by introducing the vapor mass fraction $x_v = \left(M_v/M_U\right)\chi_v$ as a measure of the species concentration. Here, χ_v gives the molar fraction of the vapor, M_v is the molar mass of the vapor and $M_U = \chi_v M_v + \chi_\text{GHe} M_\text{GHe}$ is the total molar mass of the ullage gas consisting of vapor and helium. Furthermore, the mixing of the vapor with GHe corresponds to the vapor diffusion coefficient D_v. Thus,

$$\frac{\partial x_v}{\partial t} + \mathbf{u}_U \cdot \nabla x_v = D_v \nabla^2 x_v , \tag{2.30}$$

$$\varrho_\text{ref}\left[\frac{\partial \mathbf{u}_U}{\partial t} + \left(\mathbf{u}_U \cdot \nabla\right)\mathbf{u}_U\right] = -\nabla p + \mu \nabla^2 \mathbf{u}_U + \varrho_\text{ref}\,\mathbf{g} - \varrho_\text{ref}\,\beta\left(\vartheta_U - \vartheta_\text{ref}\right)\mathbf{g}, \tag{2.31}$$

$$\varrho_\text{ref}\,c_{v,U}\left[\frac{\partial \vartheta_U}{\partial t} + \mathbf{u}_U \cdot \nabla \vartheta_U\right] = k_U \nabla^2 \vartheta_U . \tag{2.32}$$

The energy equation for the gas phase is based on an open system where gas can either enter (pressurization) or leave (venting) the tank. In a closed system, the volume expansion is caused by the phase change yielding an additional mass flux in the ullage (evaporation) or a mass reduction caused by condensation. The compression rate in the ullage can be expressed by

$$c = \frac{1}{V_U}\int_{A_\Gamma} \mathbf{u}_L \cdot \mathbf{n}_\Gamma + \frac{\varrho_L - \varrho_U}{\varrho_L\,\varrho_U}\,j\,dA . \tag{2.33}$$

Based on equation (2.6), the total mass conservation in the ullage gives

$$\nabla \cdot \mathbf{u}_U = \pm c . \tag{2.34}$$

Furthermore, the mass flux at the phase interface is

$$j = \varrho_L\left(\mathbf{u}_L\,\mathbf{n}_\Gamma - u_\Gamma\right) = \varrho_U\left(\mathbf{u}_U\,\mathbf{n}_\Gamma - u_\Gamma\right). \tag{2.35}$$

Since the non-condensable inert gas is not involved in phase change effects, the mass flux across the phase boundary is dominated by the vapor only, so that in terms of the species transport in equation (2.32), the boundary condition at the phase interface gives

$$-D_v\,\frac{\partial\left(\varrho_U\,x_v\right)}{\partial \mathbf{n}_\Gamma} = \left(1 - x_v\right)j . \tag{2.36}$$

The jump of the velocity between liquid and the gas phase can be expressed by the mass flux j, so that

$$\mathbf{u}_U = \mathbf{u}_L + \frac{\varrho_L - \varrho_U}{\varrho_L \, \varrho_U} \, j \, \mathbf{n}_\Gamma. \tag{2.37}$$

In this connection, the jump is given by its normal fraction only since the velocity in tangential direction at the phase interphase plane does not exist[2]. As well in the ullage, the temperature condition at the phase interface is defined by the saturation temperature according to the vapor partial pressure, so that

$$\vartheta = \vartheta_{\text{sat}} \left(p_v \right). \tag{2.38}$$

In order to identify the dimensionless parameters, the transport equations are scaled by appropriate characteristic quantities that are chosen as follows: The reference density is ϱ_{ref}, the reference temperature is ϑ_{ref}, the characteristic velocity is U, the characteristic length scale is L, the characteristic time is τ, the characteristic acceleration is g, the characteristic temperature difference is Θ and the characteristic mass flux is J. Derived from these quantities, the characteristic pressure is $\varrho_{\text{ref}} \, U^2$ in accordance to the stagnation pressure. In the following, the dimensionless parameters are indicated by asterisk symbols. The scaled quantities such as velocity, pressure, temperature, time, acceleration, the gradient as well as the mass flux are defined as

$$\mathbf{u}^* = \frac{\mathbf{u}}{U}, \; p^* = \frac{p}{\varrho_{\text{ref}} \, U^2}, \; \vartheta^* = \frac{\vartheta - \vartheta_{\text{ref}}}{\Theta}, \; t^* = \frac{t}{\tau}, \; \mathbf{g}^* = \frac{\mathbf{g}}{g}, \; \nabla^* = \nabla \, L, \; j^* = \frac{j}{J}. \tag{2.39}$$

In the following, the expressions in the parenthesis terms correspond to the according characteristic numbers. Thus, the governing transport relations in the liquid taken from equations (2.24) to (2.26) read

$$\nabla^* \cdot \mathbf{u}_L^* = 0, \tag{2.40}$$

$$\left[\frac{L}{\tau \, U} \right] \frac{\partial \mathbf{u}_L^*}{\partial t^*} + \left(\mathbf{u}_L^* \cdot \nabla^* \right) \mathbf{u}_L^* = -\nabla^* p^* + \left[\frac{\mu}{\varrho_{\text{ref}} \, U \, L} \right] \nabla^{*2} \mathbf{u}_L^* + \left[\frac{g \, L}{U^2} \right] \mathbf{g}^* - \left[\frac{g \, \beta \, \Theta \, L}{U^2} \right] \vartheta_L^* \mathbf{g}^*, \tag{2.41}$$

$$\left[\frac{L}{\tau \, U} \right] \frac{\partial \vartheta_L^*}{\partial t^*} + \mathbf{u}_L^* \cdot \nabla^* \vartheta_L^* = \left[\frac{k}{\varrho_{\text{ref}} \, c_{p,L} \, U \, L} \right] \nabla^{*2} \vartheta_L^* \tag{2.42}$$

with the according boundary conditions taken from equation (2.27) to (2.29) at the phase interface

$$u_\Gamma^* = \mathbf{u}^* \cdot \mathbf{n}_\Gamma - \left[\frac{J}{\varrho_{\text{ref}} \, U} \right] j^*, \tag{2.43}$$

$$\vartheta^* = \vartheta_{\text{sat}}^* \left(p_v \right). \tag{2.44}$$

Rearranging the transport relations in the ullage taken from equations (2.30) to (2.34) yield

$$\left[\frac{L}{\tau \, U} \right] \frac{\partial x_v}{\partial t^*} + \mathbf{u}_U^* \cdot \nabla^* x_v = \left[\frac{D_v}{U \, L} \right] \nabla^{*2} x_v, \tag{2.45}$$

---

[2]By definition, only forces in normal direction can act on a plane.

$$\left[\frac{L}{\tau U}\right]\frac{\partial \mathbf{u}_U^*}{\partial t} + (\mathbf{u}_U^* \cdot \nabla^*)\,\mathbf{u}_U^* = -\nabla^* p^* + \frac{\mu}{\varrho_{\text{ref}}\,U\,L}\,\nabla^{*2}\mathbf{u}_L^* + \left[\frac{g\,L}{U^2}\right]\mathbf{g}^* - \left[\frac{g\,\beta\,\Theta\,L}{U^2}\right]\vartheta_U^*\,\mathbf{g}^*, \quad (2.46)$$

$$\left[\frac{L}{\tau U}\right]\frac{\partial \vartheta_U^*}{\partial t^*} + \mathbf{u}_U^* \cdot \nabla^* \vartheta_U^* = \left[\frac{k}{\varrho_{\text{ref}}\,c_{p,U}\,U\,L}\right]\nabla^{*2}\vartheta_U^*, \quad (2.47)$$

$$\nabla^* \cdot \mathbf{u}_U^* = \pm c^* \quad (2.48)$$

with the dimensionless compression rate defined as

$$c^* = \frac{1}{V_U}\int_{A_\Gamma} \mathbf{u}_L^*\,\mathbf{n}_\Gamma + \left\{\left[\frac{\varrho_{\text{ref}}}{\varrho_U}\right] - 1\right\}\left[\frac{J}{\varrho_{\text{ref}}\,U}\right] j^*\,dA. \quad (2.49)$$

The boundary conditions of the transport equation in the ullage at the phase interface introduced in equations (2.36) to (2.38) give

$$\left[\frac{D_v}{U\,L}\right]\frac{\partial x_v}{-\partial \mathbf{n}_\Gamma} = \left[\frac{J}{\varrho_U\,U}\right](1 - x_v)\,j^*, \quad (2.50)$$

$$\mathbf{u}_U^* = \mathbf{u}_L^* + \left\{\left[\frac{\varrho_{\text{ref}}}{\varrho_U}\right] - 1\right\}\left[\frac{J}{\varrho_{\text{ref}}\,U}\right] j^*\,\mathbf{n}_\Gamma, \quad (2.51)$$

$$\vartheta^* = \vartheta_{\text{sat}}^*\,(p_v). \quad (2.52)$$

In terms of characteristic numbers, the governing equations including their boundary conditions at the phase interface can be rewritten in non-dimensional form. In the liquid, this yields

$$\nabla^* \cdot \mathbf{u}_L^* = 0, \quad (2.53)$$

$$\mathbf{St}\,\frac{\partial \mathbf{u}_L^*}{\partial t^*} + (\mathbf{u}_L^* \cdot \nabla^*)\,\mathbf{u}_L^* = -\nabla^* p^* + \frac{1}{\mathbf{Re}}\,\nabla^{*2}\mathbf{u}_L^* + \frac{\mathbf{Bo}}{\mathbf{We}}\,\mathbf{g}^* - \frac{\mathbf{Ra}}{\mathbf{Re}^2\,\mathbf{Pr}}\,\mathbf{g}^*, \quad (2.54)$$

$$\mathbf{St}\,\frac{\partial \vartheta_L^*}{\partial t^*} + \mathbf{u}_L^* \cdot \nabla^* \vartheta_L^* = \frac{1}{\mathbf{Re}\,\mathbf{Pr}}\,\nabla^{*2}\vartheta_L^* \quad (2.55)$$

with the boundary conditions for the liquid phase at the phase interface

$$u_\Gamma^* = \mathbf{u}^* \cdot \mathbf{n}_\Gamma - \frac{\mathbf{Ja}\,\mathbf{Ev}}{\mathbf{Re}\,\mathbf{Pr}}\,j^*, \quad (2.56)$$

$$\vartheta^* = \vartheta_{\text{sat}}^*\,(p_v). \quad (2.57)$$

The dimensionless governing equations in the ullage yield

$$\mathbf{St}\,\frac{\partial x_v}{\partial t^*} + \mathbf{u}_U^* \cdot \nabla^* x_v = \frac{1}{\mathbf{Re}\,\mathbf{Sc}}\,\nabla^{*2}x_v, \quad (2.58)$$

$$\mathbf{St}\,\frac{\partial \mathbf{u}_U^*}{\partial t} + (\mathbf{u}_U^* \cdot \nabla^*)\,\mathbf{u}_U^* = -\nabla^* p^* + \frac{1}{\mathbf{Re}}\,\nabla^{*2}\mathbf{u}_L^* + \frac{\mathbf{Bo}}{\mathbf{We}}\,\mathbf{g}^* - \frac{\mathbf{Ra}}{\mathbf{Re}^2\,\mathbf{Pr}}\,\mathbf{g}^*, \quad (2.59)$$

$$\mathbf{St}\,\frac{\partial \vartheta_U^*}{\partial t^*} + \mathbf{u}_U^* \cdot \nabla^* \vartheta_U^* = \frac{1}{\mathbf{Re}\,\mathbf{Pr}}\,\nabla^{*2}\vartheta_U^*, \quad (2.60)$$

$$\nabla^* \cdot \mathbf{u}_U^* = -c^* \tag{2.61}$$

with the dimensionless compression rate

$$c^* = \frac{1}{V_U} \int_{A_\Gamma} \mathbf{u}_L^* \, \mathbf{n}_\Gamma + (\mathbf{R} - 1) \, \frac{\mathbf{Ja}\,\mathbf{Ev}}{\mathbf{Re}\,\mathbf{Pr}} \, j^* \, dA \,. \tag{2.62}$$

The respective boundary conditions in the ullage at the phase interface read

$$\frac{1}{\mathbf{Re}\,\mathbf{Sc}} \, \frac{\partial x_v}{-\partial \mathbf{n}_\Gamma} = \mathbf{R} \, \frac{\mathbf{Ja}\,\mathbf{Ev}}{\mathbf{Re}\,\mathbf{Pr}} \, (1 - x_v) \, j^* \,, \tag{2.63}$$

$$\mathbf{u}_U^* = \mathbf{u}_L^* + (\mathbf{R} - 1) \, \frac{\mathbf{Ja}\,\mathbf{Ev}}{\mathbf{Re}\,\mathbf{Pr}} \, j^* \, \mathbf{n}_\Gamma \,, \tag{2.64}$$

$$\vartheta^* = \vartheta_{\mathrm{sat}}^* \, (p_v) \,. \tag{2.65}$$

According to the expressions in equations (2.53) to (2.65), the corresponding characteristic numbers are summarized in table 2.1. The selection of the characteristic quantities strongly depends on the particular fluid dynamical case. Liquid sloshing in a closed container under the impact of Earth's acceleration represents a particular fluid dynamical case where the gravity forces dominate over the surface tension forces corresponding to high BOND numbers $\mathbf{Bo} \gg 1$. The main reason for liquid sloshing in rocket propellant tanks can be traced back to variations of the external accelerations. Hereby, the periodic variation of the lateral acceleration may cause more or less oscillations. The characteristic of this motion can be scaled by the STROUHAL number. As often found in the literature [1, 27, 37], the properties of the liquid motion can be adequately scaled when the characteristic length L is defined by the tank radius, so that

Table 2.1: List of relevant characteristic numbers.

abbrev.	name	equation	forces
Bo	BOND number	$\dfrac{\varrho g L^2}{\sigma}$	$\dfrac{\text{gravity}}{\text{surface tension}}$
Ja	JACOB number	$\dfrac{c_p \Theta}{\Delta h_v}$	$\dfrac{\text{sensible heat}}{\text{latent heat}}$
Pr	PRANDTL number	$\dfrac{\mu \, c_p}{k}$	$\dfrac{\text{momentum diffusivity}}{\text{thermal diffusivity}}$
Ra	RAYLEIGH number	$\dfrac{g \, \varrho_{\mathrm{ref}} \, \beta \, \Theta \, L^3 \, c_p}{k \, \mu 0}$	$\dfrac{\text{buoyancy}}{\text{viscosity}}$
Re	REYNOLDS number	$\dfrac{\varrho_{\mathrm{ref}} \, U \, L}{\mu}$	$\dfrac{\text{inertia}}{\text{viscosity}}$
Sc	SCHMIDT number	$\dfrac{\mu}{\varrho_{\mathrm{ref}} \, D_v}$	$\dfrac{\text{momentum diffusivity (viscosity)}}{\text{mass diffusivity}}$
St	STROUHAL number	$\dfrac{L}{\tau \, U}$	$\dfrac{\text{gravity}}{\text{surface tension}}$
We	WEBER number	$\dfrac{\varrho_{\mathrm{ref}} \, U^2 \, L}{\sigma}$	$\dfrac{\text{inertia}}{\text{surface tension}}$
Ev	evaporation number	$\dfrac{\Delta h_v \, J \, L}{k \, \Theta}$	
R	density ratio	$\dfrac{\varrho_{\mathrm{ref}}}{\varrho_U}$	

$$L = R. \tag{2.66}$$

This provides an appropriate length scale, which is proportional to the wave length of the sloshing liquid in the tank assuming the lateral sloshing mode as described at the end of this chapter.

In terms of sloshing, one has to consider that the characteristic velocity U corresponds to the oscillating wave motion of the liquid at the free surface in the tank. Since the surface tension can be neglected, the fluid dynamics are characterized by the FROUDE number that is defined as

$$\mathbf{Fr} = \frac{\mathbf{We}}{\mathbf{Bo}} = \frac{U^2}{g\,R}. \tag{2.67}$$

Because liquid sloshing in a closed tank can be considered as the motion of a standing wave, the flow (fluid velocity) is equal to the group velocity multiplied by the wave amplitude/tank radius. Hence, the FROUDE number satisfies $\mathbf{Fr} \sim 1$ and therefore $[g\,L\,U^{-2}] \sim 1$. Thus, the characteristic velocity of the lateral excited sloshing system can be established by

$$U = \sqrt{g\,L}. \tag{2.68}$$

Basic fluid dynamical attributes of the liquid motion inside the tank are characterized by the REYNOLDS number $\mathbf{Re}$. For the characteristic velocity that is introduced in equation (2.68), the REYNOLDS number changes to the GALILEI number $\mathbf{Ga}$, the ratio of gravitational forces to viscous forces. In analogy to the REYNOLDS number to characterize different flow regimes, such as laminar or turbulent flow, the GALILEI number is defined as

$$\mathbf{Ga} = \frac{g\,R^3}{\nu^2} \tag{2.69}$$

where $\nu = \mu/\varrho$ is the kinematic viscosity. For different tank radii, the GALILEI number is plotted in figure 2.2 (B) on a double logarithmic scale. Again here, for the full size application, $\mathbf{Ga}$ is shown for different gravity levels ranging from $0.1g$ to $2g$ as they might occur during the rocket launch. Furthermore, the gray areas indicate the parameter range of the experiments and the full size application respectively. The GALILEI number for the full size application utilizing LH_2 appears $2-4$ decades higher than it appears for the actual experiments that are performed with LN_2. Basically, this can be traced back to the major influence of the tank radius that appears with the power of three ranging over two orders of magnitude between the experiment and the full size application. The influence of the liquid viscosity is of less significance, since ν for cryogenics is in the order of 10^{-7} m^2 s^{-1} as shown in table 3.2. However, damping implies that energy continuously decreases due to wall and bulk dissipation. In terms of these viscous effects, damping can be expressed by the GALILEI number. The logarithmic decrement Λ

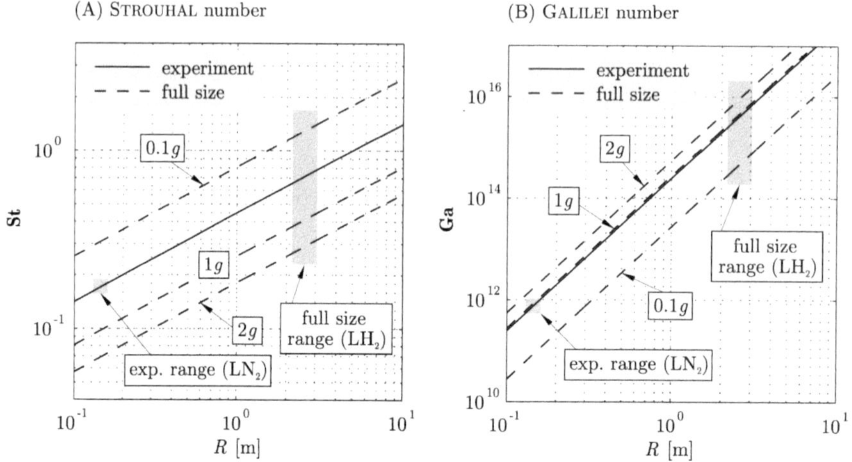

Figure 2.2: (A) STROUHAL number versus tank radius R. (B) GALILEI number versus tank radius R. Gray areas indicate the parameter range of the experiment and the full size application respectively. The excitation frequency for the full size application is assumed to be $f = 0.8$ Hz.

represents a convenient measure to determine the damping of an oscillating system. Thus, the theoretical description of liquid damping as a function of the GALILEI number **Ga** was firstly found by MILES [46] and later experimentally confirmed by STEPHENS *et al.* [55]. The relation between the logarithmic decrement and the GALILEI number is

$$\Lambda \propto \mathbf{Ga}^{-1/4}. \tag{2.70}$$

Here, the influence of the viscosity is limited to a thin boundary layer at the side walls and at the tank bottom. In this boundary layer, frictional interactions between the liquid and the wall dominate the decay of the liquid motion. This layer refers to the STOKES boundary layer [40] for an oscillating flow of a viscous fluid. The context of this relation is described within the next section in further detail.

It is assumed that the external lateral accelerations a_x and/or a_y are the main driving forces[3] to excite the liquid inside a rocket tank. This periodic oscillation is defined by the excitation frequency f and the excitation amplitude y_A. Regarding the period of one cycle, the characteristic time scale of the sloshing liquid is defined as

$$\tau = f^{-1}. \tag{2.71}$$

Thus, the STROUHAL number can be modified for gravity dominated laterally excited sloshing cases, so that

---

[3]In this work, only the lateral acceleration a_y is considered.

$$\mathbf{St} = \sqrt{\frac{R f^2}{g}} .$$ (2.72)

The STROUHAL number is plotted in figure 2.2 (A) on a double logarithmic scale to preserve the comparability between the laboratory scale experiment and the full size application. The gray areas indicate the parameter range of the respective scenarios. Of course, the STROUHAL number must be the same for the laboratory and full size conditions since τ is not constant but the variation in figure 2.2 (A) can be explained by the variation of the gravity. While the gravity level for the laboratory scale experiments is constant at $g = 9.81\ \mathrm{m\,s^{-2}}$, it may vary between $0.1\,g$ and $2\,g$ in the full size application during ascent phase. This corresponds approximately to a typical rocket launch between lift-off and reaching the orbital position since the pressure drop effect that is the subject of this investigation may occur in this period. However, despite the fact that the tank radius ranges over two orders of magnitude between experiment and the full size application, the STROUHAL number only varies slightly. Since R and g are constant, only f can be modified to adjust $\mathbf{St}$ in the current experimental case. Moreover, this might cause the change to a different sloshing mode such as the swirling mode.

Emerging from the energy balance and the boundary condition at the phase interface, the liquid properties are scaled by the PRANDTL number that is defined as

$$\mathbf{Pr} = \frac{\nu_L\, \varrho_{\mathrm{ref}}\, c_p}{k_L} = \frac{\nu_L}{D_T} .$$ (2.73)

The PRANDTL number gives the ratio of the momentum diffusivity expressed by the kinematic viscosity and the thermal diffusivity expressed by $D_T = k\,(\varrho_{\mathrm{ref}}\, c_p)^{-1}$. The thermal diffusivity is a measure of the heat transport driven by the temperature gradient within the fluid. The PRANDTL number is plotted in figure 2.3 (A) as function of the tank pressure for LN_2 as found in the experiment and LH_2 as found in the full size application. The PRANDTL number differs about a factor of two between the experiment and the full size application. Basically, this can be traced back to the thermal diffusivity, since the kinematic viscosities of the cryogenic substances appear fairly similar for the considered pressure ranges. Here, LH_2 has a higher D_T leading to smaller $\mathbf{Pr}$. Typical for cryogenic liquids, the PRANDTL number appears in the order of 1 for both, the experiment as well as the full size application. In this case, the momentum diffusivity of the fluid and its thermal diffusivity are of the same order of magnitude.

The thermodynamic properties of the fluid in particular at the liquid/vapor transition may be expressed by the JACOB number that can be extracted from the boundary conditions at the phase interface as given in equation (2.56). The JACOB number is defined by the ratio of the sensible heat to the latent heat, so that

$$\mathbf{Ja} = \frac{c_p\, \Theta_L}{\Delta h_v}$$ (2.74)

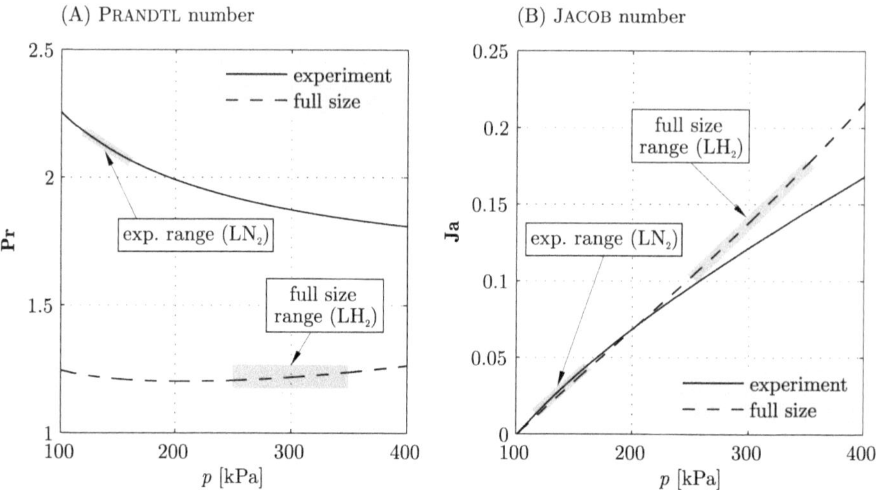

Figure 2.3: (A) PRANDTL number as a function of the tank pressure. (B) JACOB number as a function of the tank pressure. The gray ares indicate the respective parameter range for the experiments and the full size application. The LH$_2$ tank pressure for the full size application is assumed to be $p = 300$ kPa.

where Δh_v is the latent heat of vaporization. In some cases, the JACOB number also contains the density ratio ϱ_L/ϱ_U. In terms of mass transfer, the JACOB number characterizes the phase change effects in the vicinity of the free liquid surface. However, the JACOB number is plotted as function of the scaled tank pressure in figure 2.3 (B). For the experiment (LN$_2$), the JACOB number appears to be $\mathbf{Ja} < 0.05$ while the JACOB number for the full size application (LH$_2$) is in the order of $0.1 < \mathbf{Ja} < 0.175$.

Cryogenic propellant tanks in space rockets must not be considered as closed adiabatic reservoirs. In fact, external heat from the environment continuously flows into the tank provoking more or less significant thermal convection depending on the grade of insulation or the tank content, which can be liquid or vapor. This might be of particular interest regarding the common bulkhead technology that might be applied in the development of the ESC-B upper stage. The convective fluxes affect the formation of a thermal stratification due to buoyancy effects where warmer fluid in the tank lifts up. Convective flow in the fluid can be characterized by the GRASHOF number given by the ratio of the buoyancy to the viscous forces. In general, this number is defined as

$$\mathbf{Gr} = \frac{\mathbf{Ra}}{\mathbf{Pr}} = \frac{g\,\beta\,\Theta\,L^3}{\nu^2}. \tag{2.75}$$

The characteristic temperature difference Θ shall adequately scale the temperature in the tank. It is assumed that the temperature in the liquid only varies for a small extent with respect to

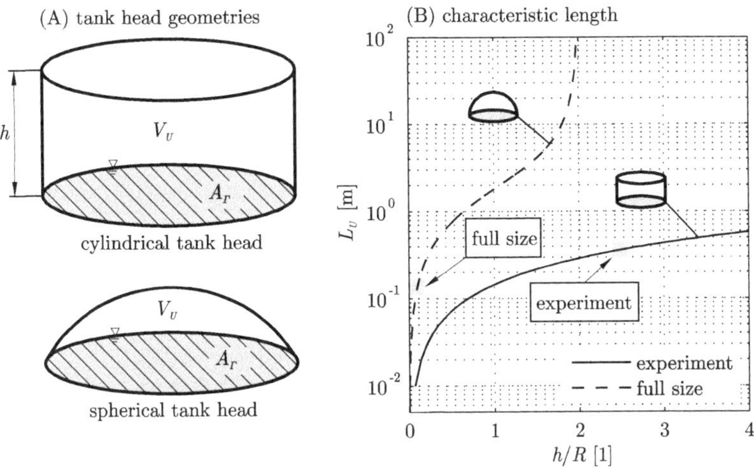

Figure 2.4: Different tank head geometries are illustrated in (A). The characteristic length L_U is plotted as function of the ullage height in (B). Here, the solid line corresponds to the experimental test case, while the dashed line corresponds to the full size application.

the temperature in the ullage that ranges over two decades. Therefore, the liquid temperature is scaled by

$$\Theta_L = \vartheta_{\mathrm{sat}} - \vartheta_{\mathrm{ref}} \tag{2.76}$$

where ϑ_{sat} is the actual saturation temperature and ϑ_{ref}. On the other side, the temperature in the ullage is scaled by

$$\Theta_U = \vartheta_{\mathrm{lid}} - \vartheta_{\mathrm{sat}} \tag{2.77}$$

with ϑ_{lid} given by the temperature of the inner wall of the tank lid.

As well the length scale L might be chosen in order to find adequate dimensions that characterize the convection effects. Therefore, the characteristic length scale in the liquid corresponds to the liquid fill level H indicating the maximum buoyancy height. Thus,

$$L_L = H \,. \tag{2.78}$$

In the ullage, the characteristic length scale may be related to the ullage size. Figure 2.4 (A) provides illustrations of different ullage shapes showing the cylindrical tank head that can be found in the experiment and the spherical tank head that can be found in the full size application (ESC-B). An appropriate measure to characterize the size of the ullage that is independent from the geometry is found by the ratio of the ullage volume V_U to the undisturbed free surface area A_Γ. Hence, the characteristic length in the ullage can be written

$$L_U = \frac{V_U}{A_\Gamma} \,. \tag{2.79}$$

The characteristic length as a function of the scaled ullage height h/R is provided in figure 2.4 (B). Here, the solid line corresponds to the cylindrical tank head, while the dashed line corresponds to the spherical tank head. The gray squares indicate the corresponding characteristic length scales. While $L_U \approx 0.1$ m is found for the full size application where the tank that is approximately 95% filled with propellant, the characteristic length in the experiment is $L_U \approx 0.3$ m for $h/R = 2.5$.

The GRASHOF number is plotted as function of the tank size expressed by the tank radius R, as shown in figure 2.5. The gray areas indicate the scope of the experiment as well as of the full size application. The GRASHOF number for the liquid is defined as

$$\mathbf{Gr}_L = \frac{g\,\beta\,(\vartheta_{\text{sat}} - \vartheta_{\text{ref}})\,H^3}{\nu_L^2} \tag{2.80}$$

and is provided in figure 2.5 (A). Assuming a larger thermal boundary layer in the full size application, the difference in $\mathbf{Gr}$ to the experiments is approximately two orders of magnitude exceeding 10^{10}. This might indicate turbulent flow in the boundary layer. In upper tank regions, the GRASHOF number for the ullage is defined as

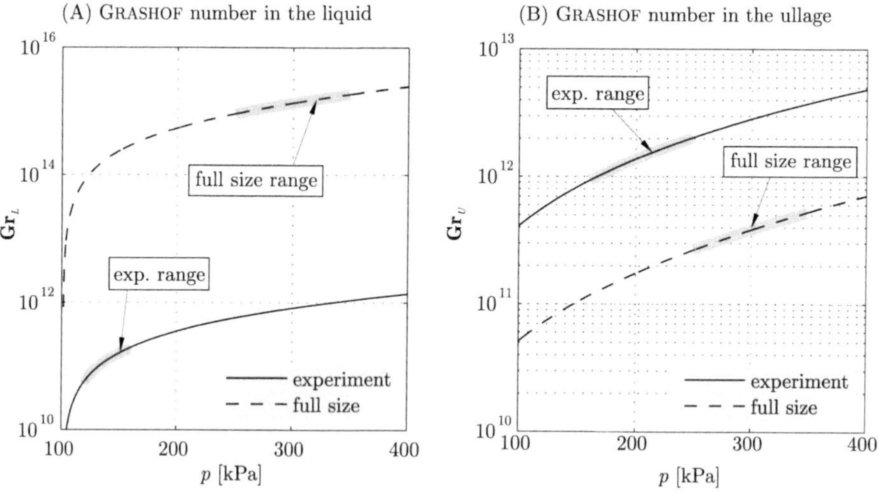

Figure 2.5: (A) GRASHOF number in the liquid region as a function of the tank pressure. The fill level for the full size application is assumed to be $H = 3$ m. (B) GRASHOF number in the ullage region as a function of the tank pressure. The gray ares indicate the according parameter range for the experiments and the full size application.

$$\mathbf{Gr}_U = \frac{g\,\beta\,(\vartheta_{\text{lid}} - \vartheta_{\text{ref}})\,V_U^3}{\nu_U^2\,A_\Gamma^3} \tag{2.81}$$

and is provided in figure 2.5 (B). Here, the difference in $\mathbf{Gr}$ between the experiment and the full size application is approximately one order of magnitude. This might vary for other fill level and therefore other ullage volumes due to the cubic impact of this parameter.

2.3 The Pressure Drop Effect

As introduced in chapter 1, under certain circumstances the sloshing motion of a cryogenic liquid in a closed reservoir may lead to a characteristic pressure drop within the tank. This might be critical compromising the structural stability of the propellant tank. A typical tank filled up to a certain level with cryogenic liquid is provided in figure 2.6 (A). The liquid temperature right after filling the propellant into the tank is indicated by (a) assuming a homogenous temperature distribution that corresponds to the saturation temperature. However, a premise to cause the characteristic pressure drop is an effective pressurization of the tank. Therefore, the tank is closed in order to allow the pressure in the tank to increase due to the external heat flowing into the reservoir provoking surface evaporation [9, 10]. This process can be accelerated by external pressurization. In both cases, the saturation temperature at the free liquid surface is coupled to the actual tank pressure. For a given saturation temperature ϑ_{sat}, the actual tank pressure p can be determined by the CLAUSIUS CLAPEYRON law or vice versa. Assuming a constant heat of vaporization Δh_v in the considered pressure range, this relation yields [54]

$$\ln\left[\frac{p}{p_0}\right] = \frac{\Delta h_v}{R_s}\left[\frac{1}{\vartheta_{\text{sat},0}} - \frac{1}{\vartheta_{\text{sat}}}\right] \tag{2.82}$$

where R_s is the specific gas constant of the ullage gas. Furthermore, p_0 and $\vartheta_{\text{sat},0}$ give a specific point on the saturation curve. Here, p_0 is the initial tank pressure after pressurization and $\vartheta_{\text{sat},0}$ is the corresponding saturation temperature.

Equation (2.82) for the experiment is provided in figure 2.6 (B) where the pressure is plotted as a function of the saturation temperature. The pressurization phase is indicated by the zone between (a) and (b) where the pressure and therefore the saturation temperature of the liquid surface increases. After reaching the required initial pressure p_0, the liquid in the tank appears in a particular thermodynamic state where the bulk is subcooled with respect to the temperature at the free liquid surface that has still saturation temperature $\vartheta_{\text{sat}}(p_0)$. Thus, the liquid at the liquid/vapor interface at the free surface is in equilibrium with the temperature distribution in the ullage where the vapor is superheated except at the free surface. The corresponding temperature distribution in the liquid is illustrated in figure 2.6 (A) showing the saturation temperature at the liquid surface. The strongest temperature gradient is assumed

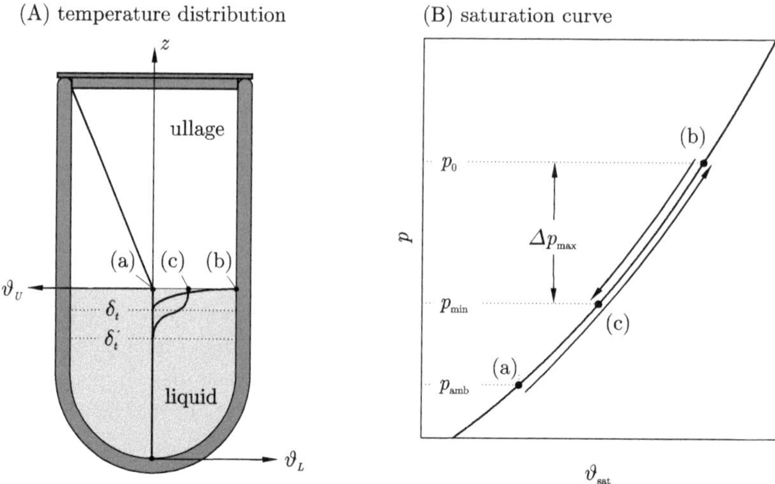

Figure 2.6: (A) Assumed temperature distribution in the liquid after filling the tank (a), after the pressurization phase (b) and after the sloshing phase (c). (B) Qualitative saturation curve given by the CLAUSIUS CLAPEYRON law [54] introduced in equation (2.82) for the equilibrium at the free liquid surface.

in the vicinity of the liquid surface where the thermal boundary layer has a depth of $\delta_t \ll R$ for the undisturbed liquid surface. Furthermore, the liquid from the bulk appears subcooled with respect to the saturation temperature at the free surface.

During the sloshing phase, the liquid from the thermal boundary layer in the vicinity of the free surface is efficiently mixed with colder liquid from the bulk below due to the flow dynamics. The saturation temperature at the free surface decreases significantly showing a more homogeneous temperature in the thermal boundary layer [39] as indicated by (c) in figure 2.6 (A). While the saturation temperature decreases, the tank pressure must decrease as well according to the CLAUSIUS CLAPEYRON relation introduced in equation (2.82). This is indicated by Δp in figure 2.6 (B). The released latent heat of condensation (evaporation) Δh_v is assumed to be dissipated through the liquid, so that the liquid temperature in average increases.

The pressure drop stops when the mixing of the liquid in the sloshing area expressed by δ_t is completed. This is indicated by (c) in figure 2.6 (B) reaching the pressure minimum $p_{\min}$. Under the impact of the sloshing liquid, the depth of the thermal boundary layer changes. This is indicated by δ_t' showing a larger distance from the free surface with a small temperature gradient as shown in figure 2.6 (A).

The impact of the pressure drop may be characterized by its magnitude and by the time dependency expressed by the pressure gradient. The pressure drop magnitude $\Delta p_{\mathrm{mag}} = p_0 - p_{\min}$

is defined as the difference between the initial pressure and the pressure minimum. This is indicated in figure 2.6 (B). The pressure gradient $\partial p / \partial t$ is a measure of the temporal response of the system. It can be considered as pressure loss per time. According to DAS & HOPFINGER [26] and HOPFINGER & DAS [36] the pressure gradient can be determined by

$$\frac{\partial p}{\partial t} = \frac{R_s}{V_U} \vartheta_{U,0} \frac{\partial m_U}{\partial t} + \frac{p_0}{\vartheta_{U,0}} \frac{\partial \vartheta_U}{\partial t} \tag{2.83}$$

with the vapor mass change rate at the phase interface $\partial m_U / \partial t = \varrho_{U,0} A_\Gamma u_v$. In this equation, only condensation effects at the free liquid surface are considered. Considering $L_U = V_U / A_\Gamma$ and rearranging yields

$$\frac{\partial p}{\partial t} = \underbrace{\frac{u_v R_s \varrho_{U,0} \vartheta_{U,0}}{L_U}}_{(I)} + \underbrace{\frac{p_0}{\vartheta_{U,0}} \frac{\partial \vartheta_U}{\partial t}}_{(II)} \ . \tag{2.84}$$

with $R_s \varrho_{U,0} \vartheta_{U,0} = p_0$. In equation (2.84), (I) gives the pressure drop fraction caused by condensation and (II) gives the pressure drop fraction caused by cooling down the ullage due to sloshing. The interfacial vapor velocity u_v under the impact of sloshing is determined by an approach by DAS & HOPFINGER [26]. They are considering a thermal boundary layer in the vicinity of the free surface of thickness δ_t. For the z direction normal to the free surface, the energy equation (2.26) can be rewritten in terms of the thermal diffusivity $D_T = k \left(\varrho_{\text{ref}} c_p \right)^{-1}$ in the liquid, so that

$$\frac{\partial \vartheta}{\partial t} + u_z \frac{\partial \vartheta}{\partial z} = D_T \frac{\partial^2 \vartheta}{\partial z^2} \ . \tag{2.85}$$

Adopted from HOPFINGER & DAS [36], the convective term in equation (2.85) can be replaced by an expression based on the gradient diffusion hypothesis commonly applied for turbulence modeling. In terms of a diffusion coefficient D'_e that characterizes the heat diffusion driven by the sloshing motion only, this term can be substituted by

$$u_z \frac{\partial \vartheta}{\partial z} = -D'_e \frac{\partial^2 \vartheta}{\partial z^2} \ . \tag{2.86}$$

Taking into account the effective diffusivity that is defined as

$$D_e = D'_e + D_T \ , \tag{2.87}$$

equation (2.85) can be rewritten

$$\frac{\partial \vartheta}{\partial t} = D_e \frac{\partial^2 \vartheta}{\partial z^2} \tag{2.88}$$

where $D_e = D_T$ applies for motionless interfaces. According to DAS & HOPFINGER [26], the boundary condition at the liquid/vapor interface at $z = H$ is given by the interfacial vapor velocity u_v that corresponds to the temperature gradient at the surface. Thus,

$$\varrho_U \, \Delta h_v \, u_v = \varrho_{\text{ref}} \, c_p \, D_e \, \frac{\partial \vartheta}{\partial z}\bigg|_{z=H} \tag{2.89}$$

where ϱ_U is the vapor density in the ullage. The temperature gradient of the right hand side in equation (2.89) is determined by integrating equation (2.88) between $z = 0$ at the tank bottom and $z = H$ at the free surface. Furthermore, the temperature gradient can be expressed by

$$\frac{\partial \vartheta}{\partial z}\bigg|_{z=H} = -\frac{\Theta_L}{\delta_t} \tag{2.90}$$

with the characteristic temperature difference in the liquid that is already introduced as $\Theta_L = \vartheta_{\text{sat}} - \vartheta_{\text{ref}}$. The boundary layer thickness is given by

$$\delta_t = 3 \, \sqrt{D_T \, t_p} + \delta_0 \tag{2.91}$$

in analogy to momentum diffusion [26], while δ_0 gives the initial thermal boundary layer thickness depending on the fill conditions and t_p is the pressurization time. Inserting equation (2.90) in equation (2.89) yields

$$u_v = -\frac{D_e}{\delta_t} \left[\frac{\varrho_{\text{ref}} \, c_p}{\varrho_U \, \Delta h_v} \Theta_L \right] \tag{2.92}$$

that gives the interfacial vapor velocity under the impact of sloshing. This relation can be expressed in terms of the JACOB number, so that

$$u_v = -\frac{D_e}{\delta_t} \, \mathbf{R} \, \mathbf{Ja}. \tag{2.93}$$

Note that this measure can be determined experimentally by knowing the corresponding pressure and temperature gradients in the vicinity of the free surface. For comparison, the effective diffusivity may be rewritten in dimensionless form satisfying [26]

$$D_e^* = \frac{D_e}{(g \, R^3)^{1/2}}. \tag{2.94}$$

Since the sloshing motion does not affect the molecular diffusion expressed by D_T, the effective diffusion during sloshing can be expressed by a dimensionless diffusion coefficient according to equation 2.87 yielding [26]

$$D_e^{\prime *} = \frac{D_e - D_T}{(g \, R^3)^{1/2}} = \frac{D_e^{\prime}}{(g \, R^3)^{1/2}}. \tag{2.95}$$

This expression enables the comparison between tanks of different size for a given sloshing motion.

2.4   Potential Theory

This section introduces the general theory that describes the sloshing motion of a lateral excited cylindrical container filled with liquid up to a certain fill level. Furthermore, the damping characteristics are added to the theory in order to consider the viscous properties of the fluid.

Surface oscillations in closed cylindrical containers can be considered as standing waves between two walls. Hereby, the nature of the surface wave is defined by the character of the excitation; axial or lateral motion. In the latter case, the magnitude of excitation yields different symmetric and asymmetric sloshing modes that are described in the following. A convenient approach to characterize the motion of a sloshing liquid is provided by the potential theory. Some assumptions are necessary to apply this theory:

(1) inviscid fluid $\mu = 0$,

(2) irrotational flow where $\nabla \times \mathbf{u} = 0$,

(3) incompressible liquid with $\varrho_L = const$,

(4) all displacements and velocities are small,

(5) gravitational forces are dominant,

(6) surface tension can be neglected,

(7) the vessel is inelastic,

(8) harmonic excitation of the form $A = A_0 \cos(\omega\,t - \varphi)$.

Assuming irrotational flow as defined in assumption (2), the velocity of the liquid can be described by the gradient of the velocity potential of the liquid in cylindrical coordinates $\Phi(r, \theta, z, t)$ [2, 16, 27, 37, 40, 46]. Then, the velocity in the liquid is defined by the potential function

$$\mathbf{u} = \nabla \Phi \tag{2.96}$$

where the nabla operator ∇ is defined in cylindrical coordinates yielding

$$\nabla = \left[\frac{\partial}{\partial r}, \frac{1}{r}\frac{\partial}{\partial \theta}, \frac{\partial}{\partial z} \right]. \tag{2.97}$$

Furthermore, regarding inviscid liquids as defined in assumption (1), the momentum balance in equation (2.13) reduces to the EULER equation since the viscous term disappears. Then, the momentum balance can be rewritten

$$\frac{\partial \mathbf{u}}{\partial t} + (\mathbf{u} \cdot \nabla)\,\mathbf{u} = -\frac{1}{\varrho_L} \nabla p + \mathbf{g}. \tag{2.98}$$

Considering EULER's approach, the only acceleration acting on the fluid is the gravitational acceleration, so that the acceleration vector yields $\mathbf{g} = (0, 0, -g)^T$ with g corresponding to Earth gravity. The convective derivative term $(\mathbf{u} \cdot \nabla)\,\mathbf{u}$ in equation (2.98) can be rewritten by means of the tensor analysis

$$(\mathbf{u} \cdot \nabla)\,\mathbf{u} = \nabla \left[\frac{\mathbf{u}^2}{2}\right] + \mathbf{u} \times (\nabla \times \mathbf{u})\,. \tag{2.99}$$

Considering assumption (4), $\mathbf{u}^2/2$ can be neglected. Thus, the first term on the right hand side gives $\nabla(\mathbf{u}^2/2) = 0$. Furthermore, assumption (2) implies irrotational flows, so that equation (2.98) simplifies

$$\frac{\partial \mathbf{u}}{\partial t} = -\frac{1}{\varrho_L}\,\nabla p + \mathbf{g}\,. \tag{2.100}$$

The conservation of mass is given in equation (2.6). By assumption (3) this equation simplifies

$$\nabla \cdot \mathbf{u} = 0\,. \tag{2.101}$$

Substituting equation (2.96) in equation (2.101) yields the LAPLACE equation, a partial differential equation of the second order, given by

$$\nabla^2\,\Phi = 0\,. \tag{2.102}$$

In cylindrical coordinates, equation (2.102) yields

$$\frac{\partial^2 \Phi}{\partial r^2} + \frac{1}{r}\frac{\partial \Phi}{\partial r} + \frac{1}{r^2}\frac{\partial^2 \Phi}{\partial \theta^2} + \frac{\partial^2 \Phi}{\partial z^2} = 0 \tag{2.103}$$

that is valid for $0 \leq r \leq R$, $0 \leq \theta \leq 2\pi$, and $0 \leq z \leq H$. To solve the differential equation (2.103), any mathematical solution must satisfy the boundary conditions at the vessel walls and at the free surface. At the tank wall, the normal velocity of the liquid is equal to the normal velocity of the wall [16]. Hence, at the tank bottom, the boundary condition is defined by

$$\left.\frac{\partial \Phi}{\partial z}\right|_{z=0} = 0 \tag{2.104}$$

as well as at the circumferential tank wall where the velocity potential in radial direction is given by

$$\left.\frac{\partial \Phi}{\partial r}\right|_{r=R} = 0\,. \tag{2.105}$$

The third boundary condition is defined by the free liquid surface. According to BAUER [16], the boundary condition at the free surface can be derived from the dynamic condition given by the unsteady form of the BERNOULLI equation. Setting equation (2.96) in equation (2.100) and replacing the acceleration vector $\mathbf{g}$ by the gravity field $\nabla\,(-g\,z)$ yields

$$\nabla \frac{\partial \Phi}{\partial t} + \frac{1}{2} \nabla \left(\nabla \Phi \right)^2 = -\frac{1}{\varrho_L} \nabla p + \nabla \left(-g\, z \right).$$ (2.106)

Integrating equation (2.106) leads to

$$\frac{\partial \Phi}{\partial t} + \frac{1}{2} \left[\left(\frac{\partial \Phi}{\partial r} \right)^2 + \frac{1}{r^2} \left(\frac{\partial \Phi}{\partial \theta} \right)^2 + \left(\frac{\partial \Phi}{\partial z} \right)^2 \right] + \frac{1}{\varrho_L} p + g\, z = 0.$$ (2.107)

In order to satisfy assumption (4), the squared and higher power velocity terms in equation (2.107) can be neglected in comparison to linear terms [2, 27]. The linearized Bernoulli equation for steady and inviscid liquid motion yields

$$\frac{\partial \Phi}{\partial t} + \frac{1}{\varrho_L} p + gz = 0.$$ (2.108)

Assuming that the liquid pressure at the liquid surface equals the static pressure of the gas in the ullage, so that the over pressure at the liquid surface $p = 0$, equation (2.108) simplifies representing the dynamic boundary condition at the free surface given by

$$\frac{\partial \Phi}{\partial t} + g\, \zeta = 0$$ (2.109)

where $\zeta = z - H$ is the small displacement of the free surface during sloshing related to the quiet liquid surface when $z = H$ (see figure 2.1). The kinematic boundary condition at the free surface is defined as relation between the surface displacement ζ and the vertical component of the liquid velocity [16], so that

$$\frac{\partial \zeta}{\partial t} = \frac{\partial \Phi}{\partial z}.$$ (2.110)

Deriving equation (2.109) and substituting in equation (2.110) leads to the linearized free surface condition

$$\frac{\partial^2 \Phi}{\partial t^2} + g\, \frac{\partial \Phi}{\partial z} = 0,$$ (2.111)

which is valid for $z = H$, $0 \leq r \leq R$, and $0 \leq \theta \leq 2\pi$.

2.4.1 Solution of the Laplace Equation

By means of the boundary conditions, the LAPLACE equation introduced in equation (2.103) can be solved using BERNOULLI's separation approach where the solution $\Phi \left(r, \theta, z, t \right)$ is defined as the product of individual functions of the cylindrical coordinates R, G, Z and the time [1], so that

$$\Phi \left(r, \theta, z, t \right) = R(r)\, G(\theta)\, Z(z)\, e^{i \omega t}$$ (2.112)

Table 2.2: Roots of the derivative of the BESSEL function of the first order $J'_m(\varepsilon_{mn}) = 0$.

		m				
		0	1	2	3	4
	0	0	-	0	0	0
	1	3.831	1.841	3.054	4.201	5.318
n	2	7.016	5.331	6.706	8.015	9.282
	3	10.173	8.536	9.969	11.346	12.682
	4	13.324	11.706	13.170	14.585	15.964

where $\omega = 2\,\pi\,f$ is the angular velocity with respect to the excitation. According to BAUER [16], a solution of the LAPLACE equation under consideration of the boundary conditions is given by

$$\Phi\left(r,\theta,z,t\right) = \sum_{m=0}^{\infty}\sum_{n=0}^{\infty} A_{mn} \underbrace{\cosh\left(\varepsilon_{mn}\frac{z}{R}\right)}_{Z(z)} \underbrace{J_m\left(\varepsilon_{mn}\frac{r}{R}\right)}_{R(r)} \underbrace{\cos\left(m\,\theta\right)}_{G(\theta)} \mathrm{e}^{i\omega_{mn}t} \tag{2.113}$$

where A_{mn} is an arbitrary constant determined from the initial condition. The wave mode is given by the mode number m describing the number of wave peaks in circumferential direction, while the wave number n describes the number of wave peaks in radial direction. The constant ε_{mn} gives the roots of the derivative of the BESSEL function of the first order J_m. An illustration of this function as well as the development of the derivatives is provided in figure 2.7 (A). According to BRONSTEIN [21], the BESSEL function of the first order is defined as

$$J_m(x) = \sum_{k=0}^{\infty} \frac{(-1)^k}{k!\,\Gamma(k+m+1)}\left(\frac{x}{2}\right)^{2k+m} \tag{2.114}$$

with $\Gamma(k)$ as Gamma function that is defined as

$$\Gamma(i) = \int_0^{\infty} x^{i-1}\,\mathrm{e}^{-x}\,dx\,. \tag{2.115}$$

The roots of $J'_m\left(\varepsilon_{mn}\right) = 0$ are provided in table 2.2 for $0 \le n \le 4$ and $0 \le m \le 4$. Contrary to ABRAMSON [1], the notation of the wave number n here starts with zero defined by the imaginary roots of the LAPLACE equation (2.112).

However, in a cylindrical tank, the velocity potentials for the individual wave modes and wave numbers are expressed by

$$\Phi_{mn} = A_{mn} \underbrace{\cosh\left(\varepsilon_{mn}\frac{z}{R}\right)}_{(I)} \underbrace{J_m\left(\varepsilon_{mn}\frac{r}{R}\right)}_{(II)} \underbrace{\cos\left(m\,\theta\right)}_{(III)} \underbrace{e^{i\omega_{mn}t}}_{(IV)}. \tag{2.116}$$

Thus, the solution of LAPLACE equation implies that sloshing is both, strongly depended on the tank size as well as on the fill level. In equation (2.116), the influence of the tank bottom is described by (I). Furthermore, the influence in radial direction is shaped by the BESSEL function as described by (II), while the influence in circumferential direction is characterized by the cosine function as described by (III). The time dependence is given by (IV).

2.4.2 Natural Frequencies

The natural frequencies of the different slosh modes ω_{mn} can be determined based on the linearized free surface condition introduced in equation (2.111). Extracting the derivatives with respect to time yields

$$\frac{\partial^2 \Phi_{mn}}{\partial t^2} = -\omega_{mn}^2 A_{mn} \cosh\left(\varepsilon_{mn}\frac{z}{R}\right) J_m\left(\varepsilon_{mn}\frac{r}{R}\right) \cos(m\,\theta)\,e^{i\omega_{mn}t} \tag{2.117}$$

as well as with respect to the vertical coordinate

$$\frac{\partial \Phi_{mn}}{\partial z} = \frac{\varepsilon_{mn}}{R} A_{mn} \sinh\left(\varepsilon_{mn}\frac{z}{R}\right) J_m\left(\varepsilon_{mn}\frac{r}{R}\right) \cos(m\,\theta)\,e^{i\omega_{mn}t}. \tag{2.118}$$

Setting equation (2.117) and equation (2.118) in equation (2.111) leads to an expression of the natural angular frequency for a given fill level $z = H$, so that

$$\omega_{mn}^2 - \frac{g}{R}\varepsilon_{mn}\tanh\left(\varepsilon_{mn}\frac{H}{R}\right). \tag{2.119}$$

Dimensionless natural frequencies are plotted versus dimensionless fill heights for different m and n as shown in figure 2.7 (B). Known from the theory [27], liquid sloshing is independent from influences emerging from the tank bottom, while exceeding fill heights $H \geq R$. This is only of limited validity for rotational symmetric tanks with certain bottom geometries. In particular, experiments have shown that spherical and convex bottom geometries indeed have strong influences on the sloshing behavior [6] even for fill level $H > R$.

The surface displacement can be determined based on the kinematic boundary condition introduced in equation (2.110). Hence,

$$\zeta_{mn} = \int \frac{\partial \Phi_{mn}}{\partial z}\bigg|_{z=H} dt. \tag{2.120}$$

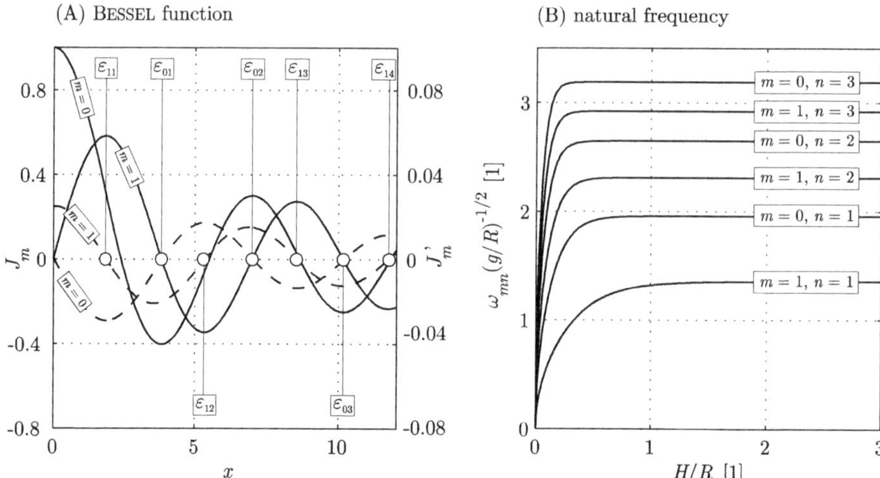

Figure 2.7: (A) BESSEL function of the first order J_m (solid lines) including its derivatives J'_m (dashed lines) for $m = 0$ and $m = 1$. The left ordinate corresponds to J_m, while the right ordinate corresponds to J'_m. Furthermore, the $\circ$ symbols correspond to the roots of J'_m indicating the maxima of J_m. (B) Dimensionless natural angular frequencies of the first sloshing modes ($m = 0, 1$, $n = 1, 2, 3$) for the scaled fill level H/R.

Setting the derivative in equation (2.118) with respect to z and solving the integral yields

$$\zeta_{mn} = -i \underbrace{\frac{1}{\omega_{mn}} \frac{\varepsilon_{mn}}{R} A_{mn} \sinh\left(\varepsilon_{mn} \frac{H}{R}\right)}_{B_{mn}} J_m\left(\varepsilon_{mn} \frac{r}{R}\right) \cos(m\theta)\, e^{i\,\omega_{mn}\,t}, \qquad (2.121)$$

so that

$$\zeta_{mn} = -i\, B_{mn}\, J_m\left(\varepsilon_{mn} \frac{r}{R}\right) \cos(m\theta)\, e^{i\,\omega_{mn}\,t}. \qquad (2.122)$$

Applying the complex law $\tilde{r}\, e^{i\tilde{\omega} t} = \tilde{r}\left(\cos(\tilde{\omega} t) + i\,\sin(\tilde{\omega} t)\right)$, the exponential term in equation (2.122) can be rewritten

$$\zeta_{mn} = -i\, B_{mn}\, J_m\left(\varepsilon_{mn} \frac{r}{R}\right) \cos(m\theta)\left[\cos(\omega_{mn} t) + i\,\sin(\omega_{mn} t)\right], \qquad (2.123)$$

and

$$\zeta_{mn} = B_{mn}\, J_m\left(\varepsilon_{mn} \frac{r}{R}\right) \cos(m\theta)\left[\sin(\omega_{mn} t) - i\,\cos(\omega_{mn} t)\right]. \qquad (2.124)$$

For visualization, only the real part of equation (2.124) is considered yielding

$$\mathrm{Re}(\zeta_{mn}) = B_{mn}\, J_m\left(\varepsilon_{mn} \frac{r}{R}\right) \cos(m\theta)\, \sin(\omega_{mn} t). \qquad (2.125)$$

Illustrations of the different wave modes for $0 \leq m \leq 3$ and $1 \leq n \leq 3$ are provided in figure 2.10 at the end of this chapter. The symmetric modes are provided in figure 2.10 (A),

(C) and (E). They are characterized by a symmetrical axial peak whereas the shape is defined by the BESSEL function of the first kind J_0. For increasing wave number n the number of waves in radial direction increases as well. Higher symmetrical modes are highly difficult to observe since their excitation frequency is too high to compensate the inertia of the liquid. Therefore, these modes appear in general as chaotic sloshing and liquid splashing where droplets may disperse from the free liquid surface.

The asymmetric modes are provided in figure 2.10 (B), (D) and (F). The asymmetric sloshing modes can be generated by axial and lateral excitation [25]. The occurrence of the sloshing mode depends only on the excitation frequency. Based on the first asymmetric mode that approximately corresponds to a flat disk tilted along its lateral middle axis[4], the asymmetric modes of higher order are characterized by asymmetric surface oscillations that are defined by a wave form in circumferential direction. Here, the number of surface waves in circumferential direction is given by the mode number m. The first asymmetric mode (lateral mode) represents a typical sloshing mode that occurs reproducible particularly for small excitation frequencies in the order of the first natural frequency.

Reaching the first natural frequency of the system, the liquid tends to turn into chaotic sloshing and the swirling mode where the lateral excited liquid wave rotates along the inner tank wall. The rotation direction may change occasionally.

2.5   Damping

The consideration of liquid sloshing in terms of the potential function Φ is only valid for irrotational and inviscid fluids. In fact, rocket propellants must be considered as viscid fluid. It is assumed that the damping properties of a fluid basically can be traced back to the presence of vortices in a small boundary layer δ between the liquid and the solid wall where energy is dissipated [37]. This energy dissipation inside the boundary layer has a much higher impact on the damping characteristics than the viscous interactions between the fluid elements. In order to satisfy the damping, the potential theory presented in the previous section is modified by regarding the momentum caused by molecular transport as well. According to equation (2.13), the momentum balance for an incompressible fluid can be written

$$\frac{\partial \mathbf{u}}{\partial t} + (\mathbf{u} \cdot \nabla)\,\mathbf{u} = -\frac{1}{\varrho}\,\nabla p + \mathbf{g} + \nu\,\nabla^2 \mathbf{u} \qquad (2.126)$$

where ν is the kinematic viscosity of the fluid. To regard the viscous properties of the fluid, the potential function in equation (2.96) is modified by adding the curl (rotation) of the vector potential function $\mathbf{A}$ in order to express the fluids viscous property. According to IBRAHIM [37], the velocity of a viscid fluid can be expressed by

---

[4]In fact, the surface is slightly curved in the vicinity of the tank wall to satisfy the contact angle criteria between liquid and wall.

$$\mathbf{u} = \nabla\Phi + \nabla \times \mathbf{A} \tag{2.127}$$

where Φ is the known velocity potential taken from the inviscid case. By substituting equation (2.127) in equation (2.126), this yields the known free surface boundary condition for the inviscid flow

$$\frac{\partial \Phi}{\partial t} = -\frac{1}{\varrho} p - gz \tag{2.128}$$

and for the viscid flow

$$\frac{\partial \mathbf{A}}{\partial t} = \nu \nabla^2 \mathbf{A} \tag{2.129}$$

that is analogous to the diffusion equation. However, for a harmonic excitation where $\mathbf{A} \propto e^{i\omega t}$, equation (2.129) takes the form

$$\left[\nabla^2 + \frac{i}{\delta_s^2} \right] = 0 \tag{2.130}$$

where $\delta_s \propto \sqrt{\nu/f}$ is defined by the STOKES boundary layer thickness at the wall. This includes that $\mathbf{A}$ can be set to zero for the liquid away from the wall. According to the energy dissipation approach by LAMB [40], the vector potential function $\mathbf{A}$ can be converted into an expression for the damping. The detailed derivation of $\mathbf{A}$ is introduced in [37] and would exceed the focus of this work.

A convenient measure that can be determined experimentally to express viscous damping represents the logarithmic decrement Λ. Further explanation about the determination of the logarithmic decrement from the experimental data are provided in chapter 4. However, the dissipation of energy can be identified at the surface, the side walls and the bottom wall. Thus, the total damping of the liquid is composed of

$$\Lambda = \Lambda_t + \Lambda_s + \Lambda_b \tag{2.131}$$

where the damping impact of the surface [37] is given by

$$\Lambda_t = 4\,\pi\,\nu\,\varepsilon_{mn} \left(R^2\,w_{mn}^2 \right)^{-1}, \tag{2.132}$$

the damping impact of the side walls is given by

$$\Lambda_s = \frac{\pi}{R} \sqrt{\frac{\nu}{2\,\omega_{mn}}} \left[\frac{1 + (n\,R/\varepsilon_{mn})}{1 - (n\,R/\varepsilon_{mn})} - \frac{2\,\varepsilon_{mn}\,H/R}{\sinh(2\,\varepsilon_{mn}\,H/R)} \right], \tag{2.133}$$

and the damping impact of the tank bottom is given by

$$\Lambda_t = \frac{\pi}{R} \sqrt{\frac{\nu}{2\,\omega_{mn}}} \frac{2\,\varepsilon_{mn}}{\sinh(2\,\varepsilon_{mn}\,H/R)}. \tag{2.134}$$

The corresponding angular frequency is taken from equation (2.119). The damping at the liquid surface is small compared to the wall damping, so that $\Lambda_t \approx 0$. Furthermore, for small

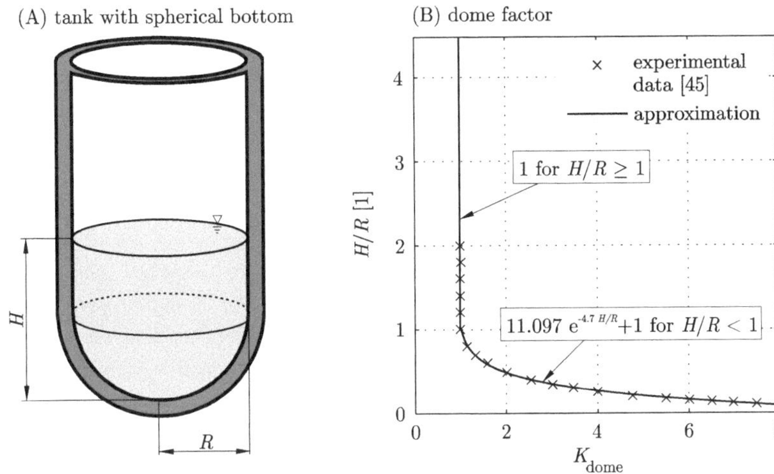

Figure 2.8: (A) Tank with spherical bottom and a fill level of $H/R = 2$. (B) Development of the dome factor for spherical bottom tanks based on Mikishev's investigations [45] represented by the $\times$ symbols. The solid line is a fit to this data.

tanks with the liquid excited in the first asymmetric sloshing mode with $m = n = 1$, the first fraction term of the parenthesis expression in equation (2.133) can be considered as one. Thus, equations (2.132) – (2.134) can be summarized for low damping [27, 37], so that

$$\Lambda = 2\,\pi\,K\,\sqrt[4]{\frac{\nu^{2}}{g\,R^{3}}}\left[\frac{1+2\,(1+2\,(1-H/R))}{\sinh\left(2\,\varepsilon_{11}\,H/R\right)\,\tanh^{1/4}\left(\varepsilon_{11}\,H/R\right)}\right]. \tag{2.135}$$

With respect to the Galilei number **Ga**, equation (2.135) can be rewritten

$$\Lambda = 2\,\pi\,K\,\mathbf{Ga}^{-1/4}\left[\frac{1+2\,(1+2\,(1-H/R))}{\sinh\left(2\,\varepsilon_{11}\,H/R\right)\,\tanh^{1/4}\left(\varepsilon_{11}\,H/R\right)}\right]. \tag{2.136}$$

The theoretical description of liquid damping as a function of the Galilei number **Ga** was firstly found by Miles [46] and later experimentally confirmed by Stephens *et al.* [55]. Here, the influence of the viscosity is limited to a thin boundary layer at the side walls and at the tank bottom. In this boundary layer, frictional interactions between the liquid and the wall dominate the decay of the liquid motion. This layer refers to the Stokes boundary layer [40] for an oscillating flow of a viscous fluid. The damping coefficient K is an experimental parameter that is defined as

$$K = K_{\mathrm{deep}}\,K_{\mathrm{dome}} \tag{2.137}$$

taking into account the bottom geometry of the tank. Hence, the damping coefficient for deep tanks K_{deep} is only dependent on the excitation and therefore on the respective sloshing mode. For the first asymmetric (lateral) mode, it is defined as

$$K_{\mathrm{deep}} = \varepsilon_{11}^{-1/4}\,. \tag{2.138}$$

As previously depicted by STEPHENS $et\ al.$ [55] as well as DODGE [27], the damping coefficient for deep tanks gives $K_{\mathrm{deep}} = 0.83$. In this configuration, K appeared to be constant for all considered fill levels $H/R \geq 1$.

For fill levels $H/R < 1$, the tank bottom geometry has to be taken into account, which provides an additional contribution to the damping characteristic of the excited liquid. For the spherical bottom geometry, MIKISHEV $et\ al.$ found that K itself is a function of the fill level for $H/R < 1$ [27, 45]. Furthermore, it was observed that the damping characteristics significantly vary when the liquid is in or slightly above the spherical dome [27]. For water, they presented a relation based on their experimental work that describes the dome factor for damping as function of the fill level [45] as shown in figure 2.8 (B). Herein, the $\times$ symbols represent their experimental data, while the solid line is an exponential fit. It gives

$$K_{\mathrm{dome}} = 11.097\,\mathrm{e}^{-4.7\,H/R} + 1\,, \tag{2.139}$$

while $K_{\mathrm{dome}} = 1$ for $H/R > 1$. Since K_{dome} is independent of the liquid properties and the tank size, equation (2.139) is valid for all cylindrical tank configurations with a spherical bottom geometry.

2.6 Spring Mass Model

In terms of the upper stage attitude control system, the liquid motion inside propellant tanks is often described by analogous models where the damping characteristic is of major importance. According to SCANLAN $et\ al.$ [53] and more recently to DODGE [27] and IBRAHIM [37], laterally excited liquid with a free surface can be considered as pendulum or spring mass system. In the latter case, the sloshing liquid is substituted by spring elements as described in figure 2.9 (A), while the viscous characteristics corresponding to the liquid damping can be considered by means of the respective dashpots. Defining the displacement of the liquid at the tank wall as $\zeta\,(t)$, the position of the liquid surface can be linked to the tank motion $y\,(t)$ [2, 6, 27] using an ordinary differential equation of second order such as

$$m_s\,\ddot{y}\,(t) = -k_s\,y\,(t) - d_s\,\dot{y}\,(t) + F_e\,(t) \tag{2.140}$$

where m_s is the sloshing liquid mass and $k_s = m_s\,\omega_{mn}^2$ gives the spring constant. In terms of the damping ratio γ, the damping rate yields $d_s = 2\,m_s\,\gamma\,\omega_{mn}$, while F_e is the excitation force given by

$$F_e\,(t) = m_s\, y_A\, \omega_{mn}^2\, \cos\,(\omega\,t)\,. \tag{2.141}$$

Here, $F_0 = m_s\, y_A\, \omega_{mn}^2$ is the initial force. Then, rearranging equation (2.140) yields

$$\ddot{y}\,(t) + 2\,\gamma\,\omega_{mn}\,\dot{y}\,(t) + \omega_{mn}^2\, y\,(t) = y_A\,\omega_{mn}^2\,\cos(\omega\,t)\,. \tag{2.142}$$

In this equation, the dominant parameter influencing the decay characteristic of the system is represented by the damping ratio γ. This parameter can be extracted from the logarithmic decrement Λ. According to [43], the damping factor is defined as

$$\gamma = \sqrt{\frac{\Lambda^2}{4\pi^2 + \Lambda^2}}\,. \tag{2.143}$$

In the literature [27, 37, 26], equation (2.143) is often simplified as $\gamma = \Lambda/2\,\pi$ for $\Lambda \ll 1$ as considered here. The damping factor is dimensionless and defined by the ratio of the damping to the critical damping. The critical damping case is defined for $\gamma = 1$ showing the typical aperiodic behavior.

However, the solution of the ordinary differential equation (2.142) is superposed by both, the homogenous solution $y_h\,(t)$ and the particular solution $y_p\,(t)$, so that

$$y\,(t) = y_h\,(t) + y_p\,(t)\,. \tag{2.144}$$

2.6.1   Decay Function

When the excitation of sloshing liquid is abruptly turned off, the decreasing motion of the system can be described by the exponential decay function. Thus, setting the right hand side of equation (2.142) equal to zero (without excitation), the homogenous part of the differential equation yields

$$\ddot{y}\,(t) + 2\,\gamma\,\omega_{mn}\,\dot{y}\,(t) + \omega^2\, y\,(t) = 0 \tag{2.145}$$

with the homogenous solution

$$y_h\,(t) = y_A\,e^{-\gamma\,\omega_{mn}\,t}\,\cos(\omega\,t - \phi_s) \tag{2.146}$$

where ϕ_s is the phase shift between the excitation and the response. The answer of the homogeneous solution decays with time due to its exponential term, so that after $t > 1/\gamma\,\omega_{mn}$ the steady state is reached where $y_h\,(t)$ is negligible.

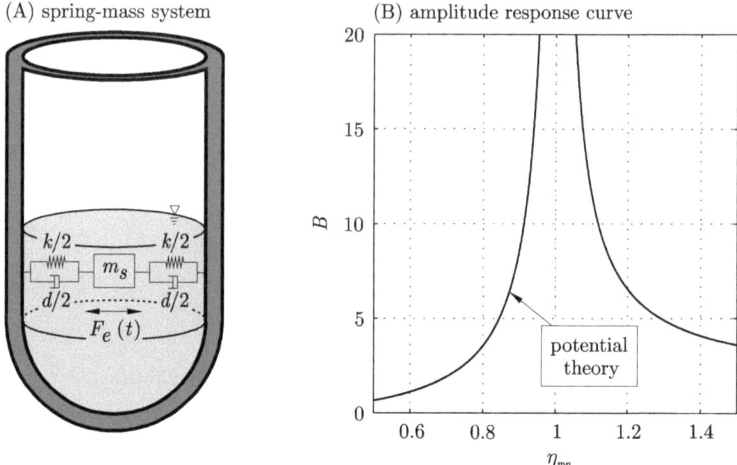

Figure 2.9: (A) Spring-mass system to describe the sloshing liquid behavior. (B) Linear response curve as function of the excitation frequency ratio η_{mn}. The solid line corresponds to equation (2.148).

2.6.2 Response Curve

For reaching the steady state, the sloshing system can be described by the particular solution of equation (2.142). This is defined by

$$y(t) = y_A\, B\, \cos(\omega\, t - \phi_s) \tag{2.147}$$

where the wave amplitude ratio $B = \zeta/y_A$ is a dimensionless measure of the wave amplitude [52] that can be determined by setting equation (2.147) as well as its derivatives in equation (2.142). Thus,

$$B = \frac{\eta_{mn}^2}{\sqrt{(1 - \eta_{mn}^2)^2 + 4\,\gamma^2\eta_{mn}^2}} \tag{2.148}$$

where

$$\eta_{mn} = \frac{\omega}{\omega_{mn}} \tag{2.149}$$

is the excitation frequency ratio [43]. Plotting B versus η gives the typical response curve, which increases for $\eta < 1$ and decreases for $\eta > 1$ as shown in figure 2.9 (B). While the liquid sloshes in phase with respect to the excitation for $\eta < 1$, the phase shift between excitation and response inverts by 180 degree. With $\phi_s + \pi$, the liquid for $\eta > 1$ sloshes contrary to the excitation. The maximum is located at $\eta = 1$ corresponding to the first natural frequency of the system. The theoretical value for the excitation frequency ratio B at $\eta = 1$ is not experimentally confirmed since the sloshing liquid in the tank changes to the rotational mode in the vicinity of the natural frequency. The swirling mode is defined by an instability where the laterally excited liquid rotates in the tank, while the direction of rotation may change alternately during the run [52]. Once reached, this mode is very stable. Therefore, the lateral sloshing mode exceeding the first natural frequency is quite difficult to realize.

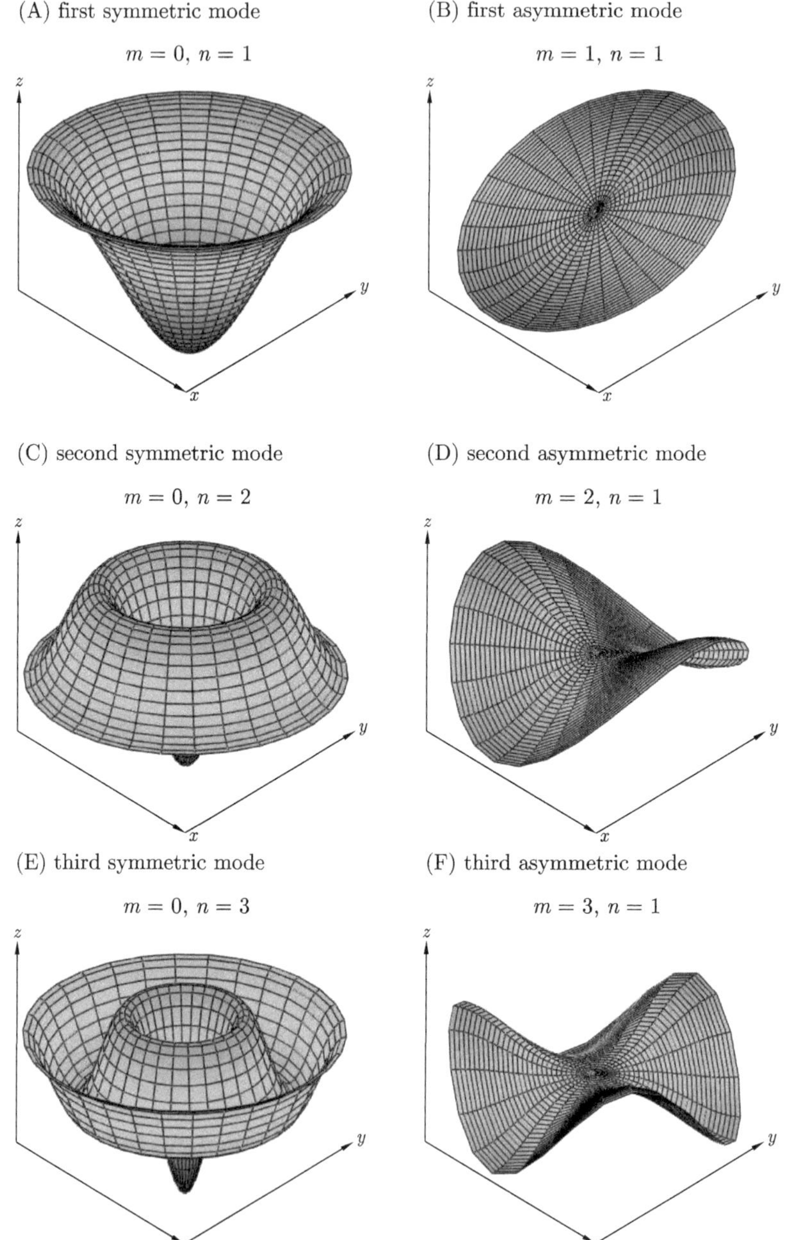

Figure 2.10: Different sloshing modes of excited liquid in a cylindrical tank. It is shown the surface contour of the sloshing liquid.

Chapter 3

EXPERIMENTAL SETUP

In this chapter, the test hardware that is required to perform the sloshing test for this work is introduced. In detail, this includes the sloshing test facility, the dewar tank and the instrumentation. The sloshing test facility was previously used by HAAKE [33], BRODA [20] and ARNDT [3, 5, 6] for lateral excited sloshing experiments with storable liquids including water and silicone oil. Some modification were necessary to enhance the facility for the utilization of cryogenic liquids such as liquid nitrogen (LN_2) affecting basically the application of an appropriate cryogenic reservoir. Furthermore, this chapter provides an overview to the equipped measurement system that is implemented in the test hardware to acquire the essential output data including sloshing forces, tank pressure and temperature distributions inside the liquid and the ullage.

3.1   Sloshing Test Facility

The principle of the test facility to perform sloshing tests with liquid nitrogen (LN_2) is basically a crank shaft design to convert the rotation of an electric motor into a linear oscillation. Figure 3.1 (A) provides an illustration of the sloshing test facility used for this work including the glass tank that contains the test liquid. Major parts of the setup are made of aluminum; the dimension of the bed plate is 2.20 m × 1.20 m. The facility is driven by the electric engine (1) with an output of 2.7 kW. This is appropriate to accelerate the given mass of approximately 200 kg. The output speed of the engine is reduced by an 1:20 transmission to ensure a smooth run. However, the rotational motion provided by the engine is converted into a linear motion by using a crank drive consisting of the eccentric (2) and the con rod (3) with length $L_c = 1.1$ m. The generated oscillation, defined by the excitation frequency $f \leq 2$ Hz and the excitation amplitude $y_A \leq 25$ mm, is approximately harmonic since $L_c \gg y_A$, so that the crank shaft ratio $\kappa_c = y_A/L_c \ll 1$. The excitation frequency is given by the engine speed, while the excitation amplitude is given by the eccentric displacement of the crank drive. Transferring the excitation, the con rod is connected to the sloshing platform (4). To allow the lateral one degree of freedom oscillation, the sloshing platform is equipped with three slide rails (5). The tank (6) containing

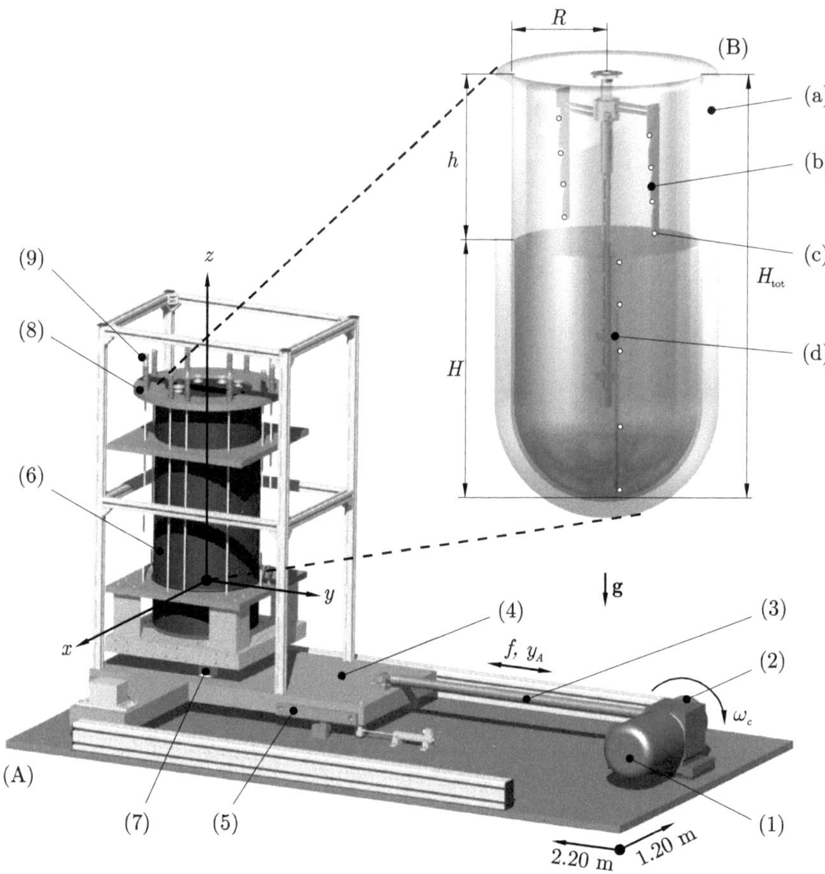

Figure 3.1: (A) Setup of the cryogenic sloshing facility. (1) engine, (2) eccentric, (3) con rod, (4) sloshing platform, (5) slide rails, (6) tank, (7) load cell, (8) lid, (9) spring. (B) The glass tank (dewar) is illustrated in detail to enable insight to the interior. (a) dewar, (b) ullage retainer, (c) temperature sensor, (d) liquid retainer.

the test liquid is installed on top of the sloshing platform. To measure the occurring sloshing forces in x, y and z direction, the tank is equipped with three load cells (7). The cylindrical tank has a radius of $R = 0.145$ m, while the bottom shape is half-spherical with a radius of $R = 0.145$ m as well. Its total height is given by $H_{\text{tot}} = 0.65$ m - 0.015 m = 0.635 m. The tank is closed by the polyacetal tank lid (8) that includes different connectors for the instrumentation and the supply. Therefore, the total tank height is reduced by 0.015 m caused by the lid reaching into the ullage. To withstand the tank pressure, the contact pressure between the lid and the glass tank is realized by 12 compression springs (9). These compression springs are also part of the security system to protect the glass tank from bursting due to over-pressurization. The critical pressure of the test facility is $p_{\text{crit}} = 180$ kPa. The fill level of the liquid is denoted as H. Except for the damping experiments where H is the variable input parameter, the fill level ist kept constant at $H = 2R$. The fill level is determined manually by using an ordinary length scale with a resolution of 1 mm. Respecting parallax effects, the accuracy for setting the fill level is $\mathcal{A} = \pm 2$ mm. Thus, the liquid volume in the tank is $V_L = 0.01543$ m^3 $\pm\, 0.00013$ m^3. The ullage height is given by $h = H_{\text{tot}} - H$. Thus, the ullage volume in the tank for a fill level of $H = 0.29$ m is $V_U = 0.02279$ m^3 $\pm\, 0.00013$ m^3.

An illustration of the tank interior is provided in figure 3.1 (B). The dewar tank (a) consists of a two layered glass container with a vacuum layer between the two glass walls to minimize the convective heat transport through the tank. Additionally, the inner glass layer is applied with a reflective coating to minimize the radiative heat losses. Details and properties of the dewar tank are provided in table A.1 as well as figure B.1 and figure B.2 in the appendix. For temperature measurement, two glass reinforced plastic retainers (b) are equipped with 4 temperature sensors each (c) to measure the temperature in the ullage. A similar retainer (d) is

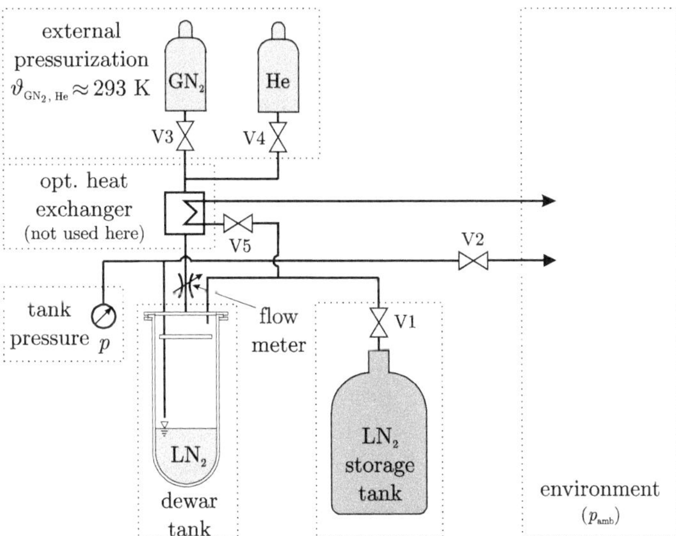

Figure 3.2: Procedural principle of the cryogenic sloshing facility including the test dewar and the storage tank for LN$_2$ and the gas bottles for GN$_2$ and GHe. The tests are conducted in the test dewar that is filled with LN$_2$ from the storage tank. Optionally, the test dewar can be externally pressurized with gaseous nitrogen or helium through the gas inlet (diameter 8 mm), while flushing is realized through the tank outlet located 5 mm above the liquid surface (diameter 8 mm). The vent line of the test tank leads outside the building into the environment.

attached beyond the liquid surface. Here, the retainer carries 5 temperature sensors to resolve the temperature stratification within the liquid. The tank pressure p is measured by using a pressure sensor located at the top of the tank. Not shown in figure 3.1 (B), the tank is equipped with a horizontal baffle located above the ullage temperature sensor retainer. The baffle enables the uniform gas propagation in the ullage during external pressurization, while preserving the thermal stratification.

The flow schematic of the test setup is illustrated in figure 3.2. Starting an experiment, the glass tank is supplied with liquid nitrogen (LN_2) from the storage tank by opening valve V1. When the desired fill level H is reached, valves V1 and V2 are closed and the tank pressure increases due to self-pressurization. The maximum vessel pressure of $p_{crit} = 180$ kPa is controlled by valve V2 for security issues. The tank can also be pressurized either with gaseous nitrogen (GN_2) or with gaseous helium (GHe) taken from external gas bottles. The gas flow rates are regulated by valves V3 and V4. At the tank lid, the flow rate is measured by a flow meter[1]. Optionally, the gas temperature can be adjusted by a heat exchanger that is fed with LN_2 from the storage tank (this part is not used here). The flow rate and therefore the temperature can be controlled by valve V5. After finishing the experimental run, the vapor from the pressurized tank is vented into the environment by using valve V2. The tank is emptied by slowly evaporating the liquid content.

3.2   Equation of Motion

The lateral one degree of freedom motion of the sloshing table is generated by a crank drive [33, 20, 3] as shown in figure 3.3. The position of the oscillating con rod tail in the pivot point (a) can be described by the quasi harmonic equation of motion that is defined as

$$y\left(t\right) + L_c = y_A \cos\left(\varphi_c\right) + L_c \cos\left(\alpha_c\right) . \tag{3.1}$$

Here, L_c gives the con rod length, while y_A is the excitation amplitude that is regulated by the eccentric displacement of the revolving point (b). Furthermore, φ_c is the rotation angle, ω_c is the angular velocity of the eccentric drive and α_c gives the deflection angle of the con rod with respect to the horizontal. The according maximum amplitudes are gained in the inner and outer dead centers that are denoted as points (c) and (d). Additionally, one has to take into account that the geometric relation describing the position of the revolving point (b) must satisfy

$$y_A \sin\left(\varphi_c\right) = L_c \sin\left(\alpha_c\right) = L_c \sqrt{1 - \cos^2\left(\alpha_c\right)}. \tag{3.2}$$

---

[1]Bronkhorst High-Tech model F-201C

In terms of the deflection angle α_c, equation (3.2) can be rewritten, so that

$$\cos\left(\alpha_c\right) = \sqrt{1 - \frac{y_A^2}{L_c^2}\sin^2\left(\varphi_c\right)}. \tag{3.3}$$

Equation (3.3) can be applied on equation (3.1) and resolved for the tank motion yielding

$$y\left(t\right) = y_A \cos\left(\omega_c t\right) + L_c \sqrt{1 - \kappa_c^2 \sin^2\left(\omega_c t\right)} - L_c \tag{3.4}$$

with the crank shaft ratio $\kappa_c = y_A/L_c$ and the rotational angle $\varphi_c = \omega_c t$. This can be rewritten as a dimensionless expression, so that

$$\frac{y\left(t\right)}{y_A} = \cos\left(\omega_c t\right) + \frac{1}{\kappa_c}\sqrt{1 - \kappa_c^2 \sin^2\left(\omega_c t\right)} - \frac{1}{\kappa_c}. \tag{3.5}$$

Here, the square root term can be approximated using the binomial series

$$\sqrt{1 - \kappa_c^2 \sin^2\left(\omega_c t\right)} = 1 - \frac{1}{2}\kappa_c^2 \sin^2(\omega_c t) - \frac{1}{8}\kappa_c^4 \sin^4(\omega_c t) - \frac{1}{16}\kappa_c^6 \sin^6(\omega_c t). \tag{3.6}$$

Experimental tests using a displacement sensor could show that the higher power terms of the binomial series can be neglected. Hence, the series is terminated after the second element and the quasi harmonic equation of motion can be rewritten

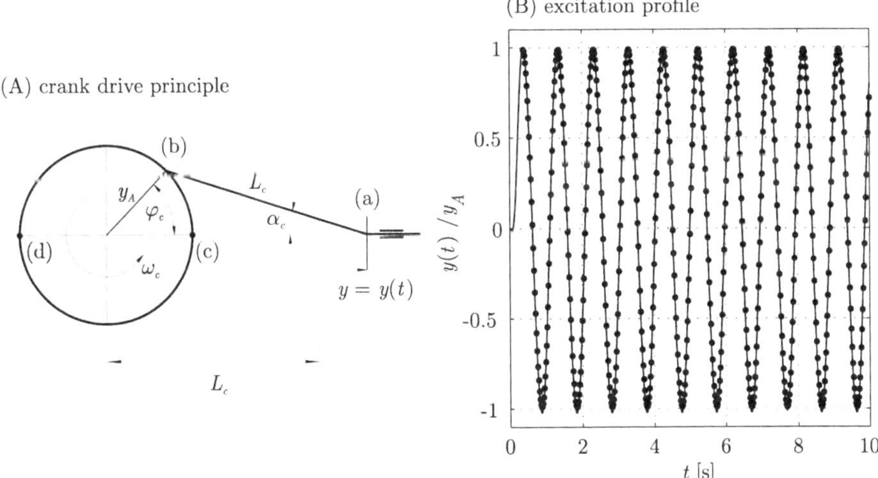

Figure 3.3: (A) Principle of the laterally excited crank drive. The pivot point is indicated by (a), while the revolving point is indicated by (b). The dead points of the eccentric are denoted (c) and (d). The sinusoidal excitation profile for $f = 1$ Hz and $y_A = 10$ mm is shown in (B) where the solid line corresponds to equation (3.7) and the full dots correspond to measurement data.

$$\frac{y\left(t\right)}{y_A} = \cos\left(\omega_c\, t\right) - \frac{1}{2}\,\kappa_c\,\sin^2(\omega_c\, t)\,. \tag{3.7}$$

The excitation profile from equation (3.7) is shown in figure 3.3 (B) for an excitation frequency of $f = 1$ Hz and an excitation amplitude of $y_A = 10$ mm. The prediction is in good agreement with the experimental data. Furthermore, the velocity is given by the time derivative of equation (3.7) yielding

$$\frac{\dot{y}\left(t\right)}{y_A} = -\omega_c\,\sin\left(\omega_c\, t\right) - \frac{1}{2}\,\kappa_c\,\omega_c\,\sin(2\,\omega_c\, t) \tag{3.8}$$

as well as the acceleration is given by the time derivative of equation (3.8) yielding

$$\frac{\ddot{y}\left(t\right)}{y_A} = -\omega_c^2\,\cos\left(\omega_c\, t\right) - \kappa_c\,\omega_c^2\,\cos(2\,\omega_c\, t)\,. \tag{3.9}$$

Regarding equations (3.7), (3.8) and (3.9), the lateral excitation of the tank is therefore fully described.

3.3   Instrumentation

Several parameters are measured prior and during the sloshing experiments to investigate the fluid-dynamical and thermodynamical behaviors of the fluid inside the tank. This includes the occurring sloshing forces, the pressure development and the temperature distribution in the ullage as well as in the liquid. The experimental data is recorded by using a LABVIEW application[2] that enables to perform the required multi-channel realtime data acquisition. Thereby, the forces and the tank pressure are measured with a rate of 10 Hz, while the temperature distribution is measured with a rate of 0.5 Hz.

3.3.1   Force Measurement

The sloshing forces during damping experiments are measured by a set of load cells[3] consisting of three single sensor devices to measure forces in x, y and z direction as shown in figure 3.4 (A) and (B). The summation of the according force component of each load cell gives the total forces in their respective directions F_x, F_y and F_z. The measurement principle is based on two quartz rings emitting electrical charges, while a load is impressed. One ring is sensitive to shear stress inducing F_x and F_y, while the other ring is sensitive to normal stress inducing F_z. The electric charges that are generated proportionally to the respective force components are led via electrodes to the charge amplifier that converts the charge signal to the corresponding force

---

[2]The measurement application is written under NATIONAL INSTRUMENTS LABVIEW 8.3
[3]The load cells are manufactured by KISTLER, Type 9366 AB, SN 534961

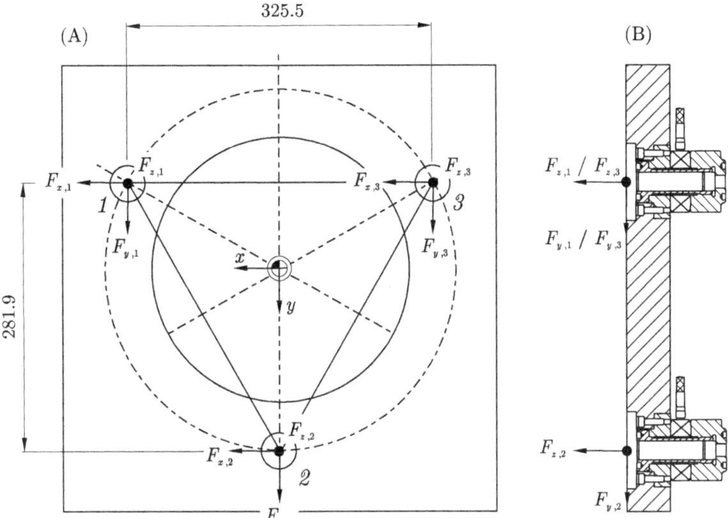

Figure 3.4: Load cells configuration to measure forces in x, y and z direction. (A) provides the top view, while (B) provides the cross-sectional area of the side view showing the load cells plane.

signal as a voltage. The resulting forces are determined by adding the singe force components for each directions. The integrated hardware low pass is applied to filter frequencies exceeding 300 Hz. The accuracy of the force measurement is determined from calibration experiments using a known test mass and a spring balance. The accuracy here is $\mathcal{A} = 1.3$ % with respect to the acquired force value

3.3.2 Pressure Measurement

The tank pressure is measured with a pressure sensor located in the tank lid[4]. The sensor is capable to resolve tank pressures in the measurement range between 10 kPa and 2500 kPa. This is in accordance to the measurement range in this work that is considered between 100 kPa and 160 kPa. By the manufacturer, the accuracy is denoted by $\mathcal{A} = 0.1$ % with respect to the acquired pressure value. The equipped pressure sensor is calibrated under ambient pressure conditions by means of a pre-calibrates pressure gauge. During the period when the experiments are conducted (approx. 4 month), the ambient pressure varied between 97.8 kPa $\leq$ $p_{\mathrm{amb}} \leq 102.0$ kPa.

---

[4]Pressure sensor TETRATEC ATM-232-0700.0147.22

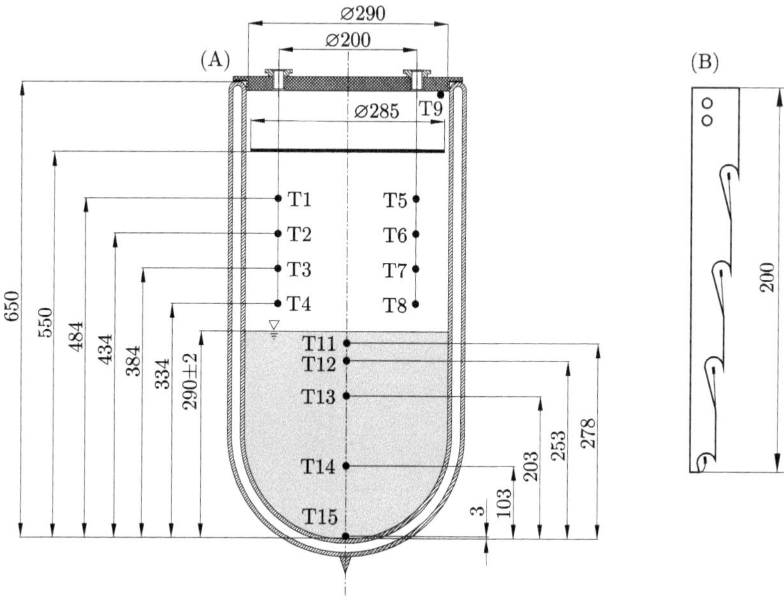

Figure 3.5: (A) Positions of the temperature sensors. All vertical dimensions are related to the origin in the center of the tank bottom. (B) Illustration of the ullage temperature sensor retainer enabling droplets easily to drain off.

3.3.3 Temperature Measurement

The temperature in the tank is measured at different locations in the ullage, in the liquid and at the wall by using silicium diodes[5] designed for cryogenic data acquisition. A total of 16 temperatures are logged by means of two serial temperature monitors[6] capable to measure 8 channels each. Specified by the manufacturer, the accuracy of the temperature sensors in cryogenic conditions between $2\,\mathrm{K} \leq \vartheta_L \leq 100\,\mathrm{K}$ is $\mathcal{A} = 0.5\,\mathrm{K}$. In the ullage, where temperatures between $100\,\mathrm{K} \leq \vartheta_U \leq 300\,\mathrm{K}$ are expected, the accuracy of the temperature sensors is $\mathcal{A} = 0.5\,\mathrm{K}$ as well. Figure 3.5 (A) provides information about the locations of the temperature sensors in the tank. Two retainers carry the ullage temperature sensors and one retainer carries the liquid temperature sensors to resolve the corresponding temperature distributions.

The ullage temperature sensor retainer is illustrated in figure 3.5 (B). When in contact with sloshing liquid from the free surface, the retainer design allows liquid droplets easily to drain off. This is supported by the small surface tension of cryogenic liquids. The retainers are fixed

---

[5]Temperature sensor DT-670 from LAKE SHORE CRYOTRONICS INC.
[6]Temperature monitor Model 218 from LAKE SHORE CRYOTRONICS INC.

Table 3.1: Groups and numbering of the temperature sensors in the tank.

location	sensor
ullage retainer I	T1, T2, T3, T4
ullage retainer II	T5, T6, T7, T8
liquid retainer	T11, T12, T13, T14, T15
inside of the tank lid	T9

$2/3\,R$ from the tank center to each side in y direction to resolve the temperature of the ullage in the y/z plane.

For the upper boundary conditions in the ullage, the temperature is measured from the inner wall at the tank lid. It is assumed that the temperature of the lid is equally distributed along the radius, so that $\vartheta_{\text{lid}} \approx 280$ K. To facilitate the readability of the figures in the next chapter, table 3.1 provides information about the numbering of the temperature sensors inside the sloshing test facility.

3.4   Heat Input

The tank used for this work is essentially a glass dewar vessel with optimized heat insulation to minimize the heat that flows into the tank. The tank walls consist of two layered glass with a vacuum layer between them and a reflective coating on the inner wall. Nevertheless, a certain amount of heat flows into the tank, so that the pressure increases continuously inside the closed system. This heat provides the thermal energy that causes evaporation inside the tank particularly at the free liquid surface. The heat that warms up the inner wall may drive convection in the tank. These effects, expressed by the GRASHOF numbers $\mathbf{Gr}_L$ and $\mathbf{Gr}_U$, are not considered in this work.

A detailed illustration of the tank is provided in figure 3.6 where the entire dewar tank system is shown in (A). The heat that flows into the tank is illustrated in (B). Here, the specific heat flux that affects the tank wall is defined as $\dot{q}_{\text{wall}}$. It is assumed that a certain amount of heat passes the vacuum that minimizes the convective heat transport in the wall. The remaining heat conducts slowly through the glass walls to reach the inside of the tank. This is similar at the tank lid that is made of polyacetal (POM). Based on the increased heat conduction ability of this material with respect to glass, the heat conducting through the lid $\dot{q}_{\text{lid}}$ also contributes to transport heat into the inner glass wall. In the ullage, the heat transport is also dominated by conduction since convective effects are small as indicated by the stratified gas. For a small extent, convection may occur at the warmer tank wall forcing warmer gas to buoy.

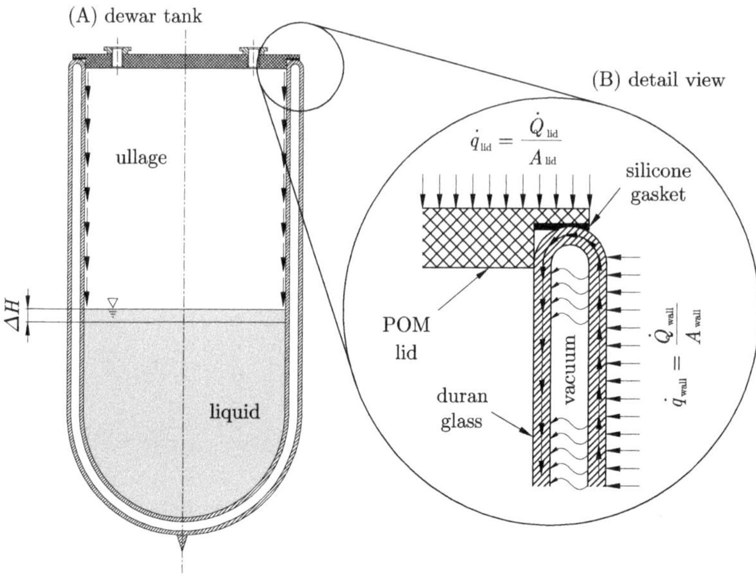

Figure 3.6: The heat flowing into the tank. The illustration in (A) shows the contact interface between the polyacetal (POM) lid and the glass tank where a silicone gasket is used; ΔH is the fill level decrease due to evaporation. The effective heat transport is indicated by the arrowed lines in (B).

Experiments including a heater element inside the tank located close to the lid could show that conductive heat transport in the ullage occurs on large time scales. These time scales are much higher than the time scales that are sufficient to explain evaporation in the liquid and therefore the instantaneous pressure increase after closing the tank.

It is assumed that some energy (radiation) is conducted through the inner tank wall, even though the inner and outer tank wall are separated by a vacuum. This is indicated by arrows in figure 3.6 (B).

However, the total amount of heat that flows into the tank is determined by measuring the decreasing fill level over a large time scale for a given fill level of $H/R = 2$. The tank was initially pre-cooled to establish experimental conditions. Due to evaporation at the free surface, the fill level reduces approximately by $\Delta H = 0.012$ m $\pm$ 0.002 m in 6 hours for an open system under ambient pressure. Therefore, the total external heat provoking vaporization can be calculated by

$$\dot{Q}_{\mathrm{heat}} = \frac{\varrho_L \, \Delta V \, \Delta h_v}{t}, \qquad (3.10)$$

Table 3.2: Fluid properties [41] corresponding to the reference temperature ϑ_{ref} given by the saturation temperature at norm pressure $p_{\text{norm}} = 101.3$ kPa for nitrogen, hydrogen and helium.

	ϑ_{ref}	$\varrho_{L,\text{ref}}$	ν_L	$c_{p,L}$	$c_{v,L}$	k_L	Δh_v	β_L	**Pr**
	[K]	[kg m^{-3}]	[m^2 s^{-1}]	[J kg^{-1} K^{-1}]	[W m^{-1} K^{-1}]	[J kg^{-1}]	[K^{-1}]		
LN$_2$	77.35	806.09	1.99E-7	2041	1084	0.15	1.99E5	0.006	2.25
LH$_2$	20.28	70.80	1.88E-7	9666	5669	0.10	4.45E5	0.017	1.24
LOX	90.16	1141.33	1.63E-7	1699	929	0.15	2.13E5	0.004	2.07

	ϑ_{ref}	$\varrho_{U,\text{ref}}$	ν_U	$c_{p,U}$	c_{vU}	k_U	β_U	**Pr**
	[K]	[kg m^{-3}]	[m^2 s^{-1}]	[J kg^{-1} K^{-1}]	[W m^{-1} K^{-1}]	[K^{-1}]	
GN$_2$	77.35	4.61	1.18E-6	1124	771	0.0075	0.015	0.81
GH$_2$	20.28	1.34	8.02E-7	12240	6603	0.017	0.064	0.77
GOX	90.16	4.45	1.47E-6	970	676	0.008	0.016	0.77
GHe	20.28	2.41	1.50E-6	5249	3121	0.026	0.050	0.72
	77.35	0.63	1.32E-5	5196	3117	0.062	0.013	0.70
	90.16	0.54	1.70E-5	5195	3117	0.069	0.011	0.69

where $\varrho_L = 806.1$ kg m^{-3} is the liquid density at norm pressure $p_{\text{norm}} = 101.3$ kPa, $\Delta V = \pi R^2 \Delta H$ is the change in liquid volume due to evaporation and $\Delta h_v - 1.99\text{E}5$ J kg^{-1} is the latent heat of vaporization for liquid nitrogen at ambient pressure. Thus, the total heat input for regarding a fill level of $H/R = 2$ is determined with the open one-species-system, so that $\dot{Q}_{\text{heat}} = 5.9$ W ± 0.5 W.

3.5 Fluid Properties

The cryogenic sloshing tests for this work are performed with liquid nitrogen (LN$_2$). This represents an appropriate approximation substituting liquid hydrogen (LH$_2$), a typical rocket propellant that is - beside liquid oxygen (LOX) - utilized in the cryogenic upper stage of the Ariane 5 space launcher. The reference properties of nitrogen, hydrogen and oxygen are provided in table 3.2. In the upper part of the table, the quantities for the cryogenic liquids are given, while the lower part of the table contains the properties for the gaseous phase. Furthermore, the properties of gaseous helium (GHe), the noncondensable pressurant gas, are

Table 3.3: Specific gas constants for the utilized fluids.

gas	GN$_2$	GH$_2$	GOX	GHe
R_s [J kg^{-1} K^{-1}]	296.8	4124	259.8	2077

listed for the corresponding saturation temperature of liquid nitrogen (LN$_2$), liquid hydrogen (LH$_2$) and liquid oxygen (LOX). The according specific gas constants for the utilized fluids are provided in table 3.3.

Chapter 4

RESULTS

This chapter presents the fluid dynamical findings as well as the thermodynamical results of the sloshing tests performed with liquid nitrogen LN_2 in a cylindrical tank with a spherical bottom geometry as shown in figure 3.1 (B). The experimental tests are performed under different considerations. Fluid dynamical properties, such as liquid damping characteristics, are analyzed by performing decay experiments in an open system. In this context, the tank content is exposed to the environment, so that the system always operates under ambient pressure conditions. On the other side, thermodynamic properties, such as pressure and temperature developments in the tank, are analyzed by sloshing tests considering a closed and pressurized system. Here, the liquid and gas properties are directly connected to the corresponding tank pressure. Furthermore, the nondimensional presentation of the experimental results allows the comparability to previous work [4, 6, 26, 39, 45, 48, 55]. At the end of this chapter, the results are summarized including predictions for the full size application based on the theory introduced in chapter 2.

Depending primarily on the excitation frequency f as well as on the excitation amplitude[1] y_A, laterally excited liquids in cylindrical tanks tend to slosh in different modes as introduced in chapter 2. To gain an impression, figure 4.1 (A) provides information about the occurring sloshing modes using LN_2 for excitation frequency ratios $0.55 \leq \eta_{11} \leq 0.9$ with respect to the first sloshing mode where $\eta_{11} = \omega/\omega_{11}$. Conducted on the test facility introduced in chapter 3, the excitation amplitude ratio is varied between $0.02 \leq y_A/R \leq 0.103$. To ensure a smooth approach of the desired excitation, the frequency is increased slowly in 0.1 Hz steps. For $0.55 \leq \eta_{11} \leq 0.8$, only the lateral mode is observed. Here, the mode number is $m = 1$ and the wave number is $n = 1$ as introduced in figure 2.10 (B). For the smallest amplitude ratios $y_A/R < 0.05$, the lateral mode appears even for approximately $\eta_{11} = 0.9$. However, for $\eta_{11} > 0.8$, the lateral sloshing mode changes to chaotic sloshing. Here, the surface motion can not be assigned to a specific sloshing mode. For further increasing excitation frequencies, the swirling mode appears. During this mode, the tank content rotates along the tank wall

---

[1]A dependency on the excitation amplitude is not considered in the sloshing theory.

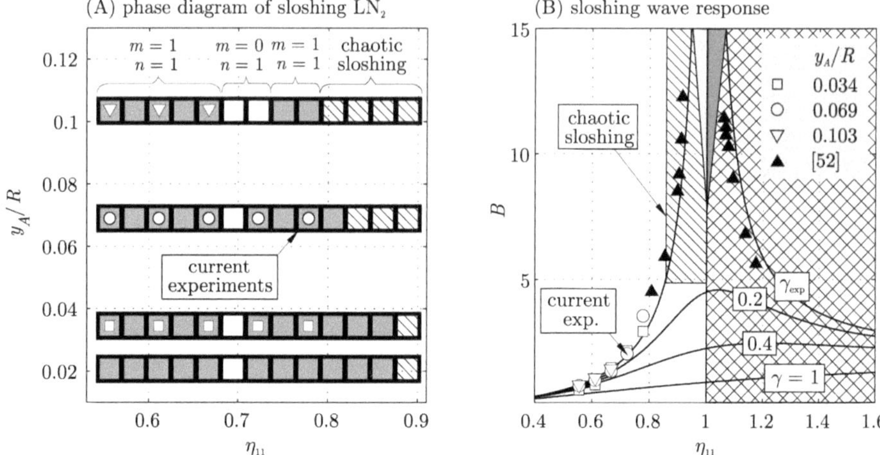

Figure 4.1: (A) Phase diagram for sloshing LN_2 in a cylindrical tank with a spherical bottom shape of size $R = 0.145$ m and a fill ratio of $H/R = 2$. The experiments are conducted on the test facility described in chapter 3. The gray areas correspond to the lateral mode where $m = 1$ and $n = 1$, while the white areas correspond to the first symmetric mode where $m = 0$ and $n = 1$. The hatched areas correspond to chaotic sloshing and the swirling mode. The current experiments are carried out for $\eta_{11} = 0.78$ and $y_A/R = 0.069$. (B) Sloshing wave response for increasing excitation frequency. The solid lines correspond to the spring mass model introduced in equation (2.148) for different γ. The full triangles correspond to data performed by ROYON et al. [52] using water in a cylindrical tank. The frequency domain where chaotic sloshing occurs in the current system using LN_2 is indicated by the hatched area. The light gray area indicates the chaotic sloshing observed by ROYON et al. [52], while the dark grey area indicates the swirling mode observed by ROYON et al. [52]. The cross-hatched area indicates frequencies that are not considered in this work.

particularly in upper liquid regions. The alternating direction of spin of the swirling mode is random. For smaller excitation amplitude ratios y_A/R the swirling mode does not appear until $\eta_{11} \geq 0.87$. This is not surprising, since for constant frequencies and decreasing amplitudes the excitation acceleration decreases as well. This leads to a more stable system particularly in the vicinity of the first natural frequency where $\eta_{11} = 1$. Furthermore for an excitation of approximately $\eta_{11} \approx 0.7$, the first symmetric sloshing mode (0,1) or axial mode is observed.

The presented experiments are carried out for $\eta_{11} = 0.78$ and $y_A/R = 0.069$ where the liquid is excited in the first asymmetric (lateral) mode (1,1). This is indicated by the marked field in figure 4.1 (A). The lateral sloshing mode provides repeatable conditions in the propellant tank where the liquid oscillates similar to a flat disk. Particularly for decay experiments, the lateral mode is appropriate to generate a steady harmonic liquid oscillation.

The corresponding sloshing wave response is provided in figure 4.1 (B) whereas the wave amplitude ratio $B = \zeta/y_A$ is plotted as a function of the frequency ratio η_{11} for different excitation amplitude ratios y_A/R. In this figure, the solid lines correspond to the wave amplitude response of the spring mass system introduced in equation (2.148) in section 2.6 for different damping factors γ. Here, the wave amplitude ratio B increases for increasing η_{11} when $\eta_{11} < 1$. For deep fill ratios $H/R > 1$ and finite excitation amplitudes, the first natural frequency of the system is reached close before reaching $\eta_{11} = 1$ [47, 52]. Exceeding the resonance for $\eta_{11} > 1$ the wave amplitude re-decreases. The experimental data using LN_2 for different amplitude ratios is indicated by the open symbols, while the full symbols correspond to data generated by Royon et al. [52] using water in a cylindrical tank. The hatched area in figure 4.1 (B) indicates chaotic sloshing. Contrary to the results presented in [52], chaotic sloshing is already observed for frequency ratios $\eta_{11} > 0.8$. This can be explained by the utilization of LN_2 having a very low dynamic viscosity. For an excitation of $\eta_{11} = 0.78$ the deflection of the free surface with respect to the horizontal gives $\alpha_s = 14.7 \pm 1.1$ degrees. Assuming an approximately flat surface[2], the deflection angle is determined by the wave amplitude ratio and the excitation amplitude ratio yielding

$$\tan \alpha_s = B\, y_a/R\,. \tag{4.1}$$

In matters of the full size application, this represents a worst case scenario that may be critical compromising the structural stability of the tank.

4.1 Experimental Matrix

All experiments that are conducted for this work are listed in table 4.1. As mentioned before, two kinds of experiments are performed: damping experiments to characterize the dynamics of the fluid as well as pressure drop experiments.

The excitation for all experiments (damping as well as pressure drop) is set to a frequency ratio of $\eta_{11} = 0.78$ and an amplitude ratio of $y_A/R = 0.069$ corresponding to an excitation frequency of $f = 1.4$ Hz and an excitation amplitude of $y_A = 10$ mm. This is appropriate to excite the liquid at the first asymmetric sloshing mode or lateral mode as it is assumed to occur in upper stage tanks during lift-off as well.

---

[2]This is only an approximation to determine the defection angle. In fact, the shape of the surface depends on the contact angle between liquid and tank wall

Table 4.1: Experiments performed for this work.

	category	variable parameter							
damping	decay exp.	fill level H/R							
		0.5	1.0	1.5	2.0	2.5			
pressure drop		initial pressure p_0 [kPa]							
	self-press.	120	130	140	150	160			
	ext. nitrogen press	120	130	140	150	160			
		helium concentration χ_{GHe} [mol mol^{-1}]							
	ext. helium press	0.35	0.46	0.52	0.6	0.68	0.73	0.79	0.87

4.2 Damping Characteristics

Information about the damping characteristics is of major importance since liquid propellant in general represents the highest weight fraction influencing the rockets attitude control system during the flight. Decay experiments represent a convenient method to determine the damping characteristics of a liquid propellant in storage tanks of space rockets [27, 30].

For the purpose of achieving a periodic initial condition, the liquid is excited in the lateral mode with a frequency in the vicinity of the first natural frequency where the frequency ratio is $\eta_{11} = 1$. Again, the excitation here is $\eta_{11} = 0.78$ and $y_A/R = 0.069$. All experiments are conducted in a cylindrical tank with a spherical bottom geometry. The sloshing liquid must reach the steady state before the excitation can be stopped; after that the oscillation decays. Typically, the steady state is reached after approximately 300 seconds. To determine the damping characteristics, it is necessary that the liquid sloshes in the first asymmetric sloshing mode where the liquid surface oscillates similar to a flat disk. In higher modes, the liquid tends to chaotic response that generally ends up in the swirling mode, which is dominated by nonlinear effects [6, 37, 47, 52].

As introduced in chapter 2, viscous liquid damping is determined by means of the logarithmic decrement Λ. The logarithmic decrement can be extracted from decay experiments by measuring the occurring sloshing force. In general, the logarithmic decrement is defined as

$$\Lambda = \ln\left[\frac{F_{i-1}}{F_i}\right] = \frac{1}{i}\ln\left[\frac{F_0}{F_i}\right] \tag{4.2}$$

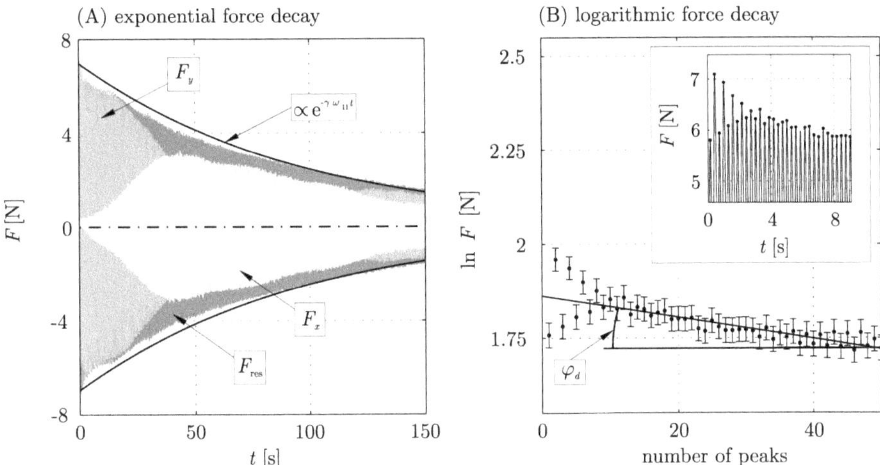

Figure 4.2: (A) Typical logarithmic decay function of a LN_2 sloshing experiment in a cylindrical tank with spherical bottom geometry. The gray shades indicate the decay of the forces in F_x and F_y as well as the resulting force F_{res}. The solid envelope curve corresponds to equation (4.5). (B) Exponential decay function regarding the first 50 oscillations. The straight solid line correspond to a linear regression to fit the data. The time domain of the first periods is highlighted in the inset indicating the non-linear behavior during the first oscillations.

where F_0 corresponds to the maximum force that occurs during the first wave amplitude and F_i gives the maximum force during the ith wave amplitude. A typical damping experiment performed for this work is represented in figure 4.2 (A) showing the decay of the sloshing force due to viscous damping in the liquid. This figure provides the decay curves for the resulting force

$$F_{res} = \left[F_x^2 + F_y^2 \right]^{1/2} \tag{4.3}$$

where F_x is the force perpendicular to the excitation direction and F_y is the force in excitation direction. The sloshing force in x direction is considered as well, since the liquid content tends to perform a slow rotation inside the tank. The reason for this slow rotation can be traced back to a small asymmetry and is not to be mistaken with the swirling mode. However, the resulting sloshing force shows the behavior of a typical exponential decay as indicated in figure 4.2 (B).

According to equation (4.2), the logarithmic decrement Λ can be determined from the maximum force amplitudes F_i of the decay experiment. This is shown in figure 4.2 (B), where the first 50 resulting force maxima and minima are plotted on a logarithmic scale. The resulting force development in the time domain during the first seconds is provided in the inset of figure 4.2 (B). Both amplitudes, in positive direction as well as in negative direction are shown. The variation of the development during the first five oscillations can be explained by a certain nonlinear behavior of the sloshing liquid [43]. Previously, this behavior was observed during decay experiments using water in an acrylic glass tank [3]. Averaging the minimum and maximum force amplitudes indicated by the solid regression line in figure 4.2 (B) allows the assumption that these nonlinear effects may compensate each other in order to be neglected. However, with respect to the angle between the regression line with the horizontal, the logarithmic decrement is determined by the trigonometric relation given by

$$\Lambda = \tan\left(\varphi_d\right) \tag{4.4}$$

where φ_d corresponds to the slope of the linear regression line.

Knowing the damping of the excited system, this enables the formulation of the decay function corresponding to the envelope curve in figure 4.2 (A). The envelope curve to describe the force decay can be extracted from the homogeneous solution of the ordinary differential equation introduced in equation (2.145). Regarding $F = m_s \omega_{11} y$ and $F_0 = m_s \omega_{11} y_A$, the solution yields an expression for the exponential decay envelope curve that is defined as

$$F = F_0 \, e^{-\gamma \omega_{11} t} \, . \tag{4.5}$$

The cosine term in equation (2.146) equals zero since the excitation is stopped during the decay. The sloshing liquid mass m_s is eliminated in equation 4.5. This parameter can be determined either experimentally from the force decay or theoretically by using a pendulum model [27]. The initial force F_0 is the average maximum sloshing force of the excited steady state case. In good approximation, F_0 can be determined from average of min/max values of the first oscillation right after stopping the excitation; the extraction of the sloshing forces from the excited case would be quite inaccurate including the measurement of the rigid body as well. Thus, the initial force gives

$$F_0 = \frac{|F_{0,\,\mathrm{min}}| + |F_{0,\,\mathrm{max}}|}{2} \tag{4.6}$$

to compensate the error during the first oscillations due to the non-linear behavior.

The parameter γ in equation (4.5) is defined as damping ratio and can be extracted from the logarithmic decrement Λ considering only underdamped systems where $\eta_{11} < 1$. As introduced in equation (2.143), the damping factor is defined as

$$\gamma = \sqrt{\frac{\Lambda^2}{4\pi^2 + \Lambda^2}} \, . \tag{4.7}$$

For the decaying LN_2 sloshing, values for Λ and γ respectively are provided in table 4.2. According to the general sloshing theory, the values for $H/R > 1$ are approximately constant. The damping for the shallow fill levels $H/R < 1$ is higher for some extent, while the value for $H/R = 1$ is smaller than expected.

Table 4.2: Logarithmic decrement and damping ratio for different fill level.

H/R	Λ	γ
0.5	1.8E-2	2.9E-3 $\pm$ 8.7E-4
1.0	4.7E-3	7.5E-4 $\pm$ 1.4E-4
1.5	6.1E-3	9.7E-4 $\pm$ 1.7E-4
2.0	5.7E-3	9.0E-4 $\pm$ 9.1E-5
2.5	4.9E-3	7.8E-4 $\pm$ 7.2E-5

The application of the potential theory on the sloshing case predicts an influence of the fill level. Furthermore, in tanks where the fill level is $H/R < 1$, the significant influence of the tank bottom geometry on the damping characteristic of the excited liquid has to be taken into account as well. Considering the spherical bottom geometry, the contribution to the viscous dissipation from the tank bottom is obtained by combining equation (2.136) and equation (2.139). Thus,

$$\gamma = 0.83 \left[11.097\, e^{-4.7\, H/R} + 1 \right] \mathbf{Ga}^{-1/4} \left[\frac{1 + 2\, (1 + 2\, (1 - H/R))}{\sinh\left(2\, \varepsilon_{11}\, H/R\right) \tanh^{1/4}\left(\varepsilon_{11}\, H/R\right)} \right], \qquad (4.8)$$

while the GALILEI number for the current experiments is $\mathbf{Ga} = 7.55\text{E}11$. The results for the damping characteristics in a tank that is partly filled with LN$_2$ are provided in figure 4.3 where the damping ratio γ is plotted as a function of the fill level H/R. Here, the $\square$ symbols correspond to the experimental LN$_2$ results, while the solid line corresponds to the theoretical prediction based on equation (4.8).

As mentioned before, the damping behavior changes for $H/R \leq 1$ showing increasing damping ratios. The increase of γ is basically related to the enhanced contact to the tank wall with respect to the total liquid volume that of course decreases for smaller H/R. The experimental results are in good agreement to the theoretical prediction particularly for deeper fill level where $H/R > 1$. Nevertheless, the system is highly sensitive due to the very low viscosity of cryogenic liquids ($\nu = 1.99\text{E-}7\ \mathrm{m^2\,s}$) particularly for low fill levels. This explains the higher error bars of γ determined for $H/R < 1$ as shown in figure 4.3.

However, unusual behavior is observed for $H/R = 1$ where the liquid fill level appears on level of the interface between the cylindrical part and the spherical part of the tank. In this case, the sloshing motion is supported by a smooth transition leading to smaller damping. This behavior was also observed during pre-tests using water and in [6].

The predicted approximation for the damping ratio using liquid hydrogen (LH$_2$) is indicated by the dotted line in figure 4.3. Obviously, the damping characteristic varies only for a small

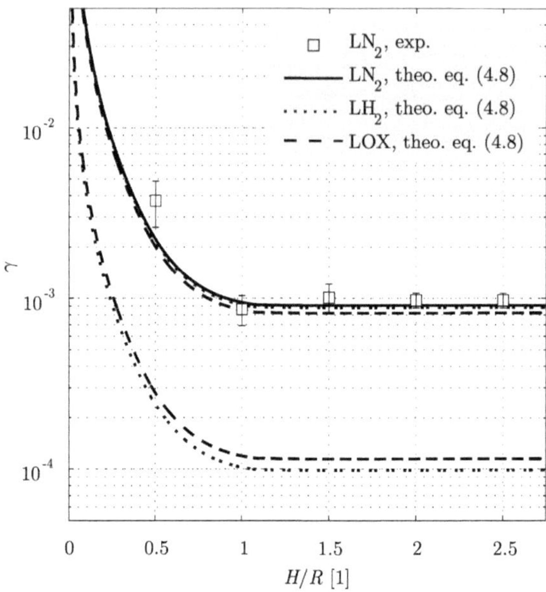

Figure 4.3: Development of the damping characteristics as function of the fill level. The ordinate is plotted on a logarithmic scale to enhance the readability. The theoretical predictions for LN$_2$, LH$_2$ and LOX are based on equation (4.8).

extent from the γ determined from the LN$_2$ experiments (solid line). This can be explained by the similarity of the kinematic viscosities of both liquids. However, the damping changes by a large extent by considering a cylindrical full size tank of radius $R = 2.5$ m with a spherical bottom geometry (e.g. ESC-A LOX tank geometry). This is indicated by the dotted line in figure 4.3 (A). Hence, the GALILEI number changes leading to smaller damping. For large tanks, the impact of the viscous friction in the STOKES boundary layer decreases, correspondingly the damping decreases as well.

The lower solid lines in figure 4.3 correspond to the theoretical prediction considering a full size LH$_2$ or LOX system. On LH$_2$ side, a tank size of $R = 2.7$ m is assumed, while the tank size on LOX side is $R = 2.0$ m. Due to the size, the damping ratio decreases by approximately one decade.

4.3   The Pressure Drop Effect

While the focus of the previous section is more targeted on the fluid dynamics of sloshing and in particular on the damping characteristics of cryogenics excited in a cylindrical tank, this section presents the thermodynamic results that are acquired by means of cryogenic sloshing tests using liquid nitrogen (LN_2). Investigations under thermodynamic consideration are of major importance since cryogenic propellants are frequently utilized in upper stage engines. Under some circumstances, a characteristic pressure drop can occur inside the upper stage hydrogen tank as a result of liquid sloshing. Pressure drops may be critical compromising the structural stability of the tank as well as challenging the pressurization system. Therefore, the understanding of these pressure drop effects may help to enhance the development of future upper stage designs.

The present investigation is focussed on the impact of thermodynamic interactions (condensation) at the phase interface between liquid and vapor under the influence of liquid sloshing. Due to their particular properties, such as boiling temperatures below 120 K in combination with remarkable low viscosity values, cryogenic fuels represent a certain species of liquid propellants, which are exceptionally sensitive to thermal influences. In order to investigate the occurring pressure drop effects, the pressure as well as the fluid temperature are measured at different locations within the tank, as introduced in chapter 3.

The experimental procedure includes the pre-cooling process of the tank that is initiated 12 hours prior to perform the actual sloshing experiments. Therefore, the tank is filled approximately to its half ($H/R \approx 2$) with liquid nitrogen (LN_2). The aim of the pre-cooling process is to ensure repeatable initial conditions inside the tank by dissipating the heat that is stored in the inner glass wall of the tank. While the tank is filled with LN_2, the heat that flows from the walls into the fluid particularly in regions where the wall is in contact with the cold liquid.

Each experiment starts by adjusting the required fill level H in the pre-cooled tank by feeding LN_2 from the reservoir into the open tank. In terms of the tank radius R, the required fill level is $H/R = 2$. While the experiments are conducted in a cylindrical tank with a spherical bottom geometry as shown in figure 3.1, this fill level is appropriate to exclude influences from the bottom shape on the sloshing behavior of the liquid. The positions of the temperature sensors in the tank are fixed. Therefore, the temperature measurement in the liquid and in the ullage prevents any fill level variations.

Repeatability studies have revealed that the system is highly sensitive to the filling process particularly to the period during, which the tank lid is removed and heat can enter unresisted to the inside. Here, heat from the environment warms up the vapor in the ullage particularly in higher strata. This effect is reduced by a delay of approximately two minutes after closing the tank lid before starting the experimental run. In this period, the venting valve is open to

Table 4.3: Dimensional p_0 and dimensionless initial pressures p_0^*. The dimensionless initial pressure is based on the thermodynamic pressure.

p_0 [kPa]	120±0.2	130±0.2	140±0.2	150±0.2	160±0.2
$p_0^* = p_0/(\varrho_{U,\mathrm{ref}}\, R_s\, \vartheta_{\mathrm{ref}})$	1.13	1.23	1.32	1.42	1.51

suppress the pressurization of the system. This is sufficient to allow repeatable temperature stratifications in the ullage.

As described in the introduction, typical rocket tanks loaded with cryogenic propellant are pressurized up to a certain tank pressure. Among others, this is also of importance to ensure the required structural stability of the tank. For the full size application, the initial pressure that is required to ensure appropriate operational conditions strongly depends on the engine design and of course on the utilized liquid propellant.

Table 4.3 provides information about the initial pressure p_0 and the respective nondimensional values p_0^* based on the scaling presented in chapter 2. The experiments are carried out between 120 kPa $\leq p_0 \leq$ 160 kPa except the experiments using gaseous helium (GHe) that are entirely performed at $p_0 = 140$ kPa. However, the tank pressurization is realized by different methods. In this work, three different scenarios are considered where the tank is pressurized by:

1. **Self-pressurization**. In this scenario, the nitrogen tank is pressurized due to evaporation effects at the free liquid surface caused by the heat that flows into the tank by conducting through the glass walls and the tank lid.

2. **GN$_2$ pressurization**. This represents a one-species system where gaseous nitrogen (GN$_2$) taken from an external reservoir is used to pressurize the tank. Here, the pressurization time scale is approximately 10 times smaller than during self-pressurization.

3. **GHe pressurization**. In this two-species system to quickly adjust the tank pressure, the nitrogen tank is pressurized with gaseous helium (GHe) that represents a non-condensable inert gas.

When the required initial pressure p_0 is reached, the lateral excitation is initiated in order to provoke liquid sloshing as it typically occurs during the ascent phase due to specific flight maneuvers. In case of external pressurization, the gas feeding is interrupted prior to start the tank motion.

However, after the required initial pressure is reached, the liquid in the tank appears in a particular thermodynamic state where the bulk liquid is subcooled with respect to the temperature at the free liquid surface that has saturation temperature $\vartheta_{\mathrm{sat}}^*$. This can be expressed by the degree of subcooling ϖ_{sub}. The liquid at the liquid/vapor interface at the free surface is in equilibrium with the ullage stratum above. Thus, a high temperature gradient forms in the liquid layer close to the surface. Under the liquid surface, the depth of the thermal boundary layer is given by δ_t. The size of the thermal boundary layer is in the order of centimeters. With the temperature instrumentation used in this work, δ_t can not be exactly determined experimentally since the temperature sensor resolution in this region is insufficient to resolve such a small range. Above the liquid/vapor interface, the temperature in the ullage increases linearly along the z-axis.

The conditions change after liquid sloshing is initiated. The oscillating free liquid surface affects mixing effects in the surface area and to a significant decrease of the surface temperature ϑ_{sat}^*. Here, the deflection of the free liquid surface α_s is sufficient to significantly influence the coupling between the heat and mass transfer. Because the temperature decreases due to mixing effects at the liquid/vapor interface, the pressure in the tank rapidly decreases in accordance to the equilibrium condition (CLAUSIUS CLAPEYRON) introduced in equation (2.82) causing condensation. This ends up in the characteristic pressure drop. Thereby, the released latent heat of vaporization Δh_v flows into the liquid. After sloshing, the liquid temperature in the thermal boundary layer shows a significant smaller temperature gradient as it appears after pressurization.

Performing numerical simulations, LACAPERE *et al.* [39] observed a strong temperature gradient in the thermal boundary layer below the free liquid surface. Due to the mixing, the temperature below the free surface increases, while the temperature at the free surface decreases to cause the pressure drop.

The condensation and therefore the pressure drop stops when the system reaches a certain minimum pressure, where the temperature in the thermal boundary layer is almost uniform. The liquid keeps sloshing in the tank, but this no longer affects the saturation temperature at the free liquid surface. From here, the tank pressure re-increases due to the heat flowing into the tank.

For data acquisition, the tank pressure as well as the temperatures at different locations in the liquid and in the ullage are measured. Pressure data is logged with a rate of 100 Hz, while the temperature is logged with a rate of 0.5 Hz limited by the inertia of the equipped temperature sensors. At the free surface, the temperature always has saturation temperature ϑ_{sat}. The saturation temperature is calculated using equation (2.82), while the tank pressure is logged by a pressure sensor located in the tank lid as described in section 3.3.2.

4.3.1 Data Scaling

As introduced in chapter 2, the data presented in this work is nondimensionally scaled in order to ensure comparability to previous results using LH_2 and LOX as far as to enable a prediction for the full size application. According to equation (2.39) and (2.76) in section 2.2, the scaled temperature in the liquid is defined as

$$\vartheta_L^* = \frac{\vartheta_L - \vartheta_{\mathrm{ref}}}{\vartheta_{\mathrm{sat},0} - \vartheta_{\mathrm{ref}}} \, . \tag{4.9}$$

The reference temperature is given in table 3.2 in the previous chapter as well as in table A.8 in the appendix accordingly. For liquid nitrogen, this yields $\vartheta_{\mathrm{ref}} = 77.35$ K. For data scaling, the liquid temperature is scaled between 0 and 1, while $\vartheta_L^* = 0$ corresponds to ϑ_{ref} and $\vartheta_L^* = 1$ corresponds to the saturation temperature $\vartheta_{\mathrm{sat},0}$ reaching the initial pressure p_0. This is of advantage, since all liquid temperature distributions merge into $\vartheta_L^* = 1$ assuming saturation conditions at the liquid surface for a given tank pressure. Quantities for $\vartheta_{\mathrm{sat},0}$ are provided in table A.8 in the appendix.

The temperature in the ullage is scaled separately since the temperature ranges in the liquid and in the ullage are too different as to apply a combined scaling concept. Therefore, the scaled ullage temperature is defined as

$$\vartheta_U^* = \frac{\vartheta_U - \vartheta_{\mathrm{sat},0}}{\vartheta_{\mathrm{lid}} - \vartheta_{\mathrm{sat},0}} \tag{4.10}$$

where $\vartheta_{\mathrm{lid}} = 280$ K $\pm$ 3 K is the inner lid temperature. The ullage temperature scale ranges between ϑ_{sat} and ϑ_{lid}. As well as in the liquid, the temperature in the ullage is scaled between 0 and 1 so that $\vartheta_U^* = 0$ corresponds to the saturation temperature of the liquid surface $\vartheta_{\mathrm{sat},0}$ for the initial pressure p_0 and $\vartheta_U^* = 1$ corresponds to the inner lid temperature ϑ_{lid}. Quantities for $\vartheta_{\mathrm{sat},0}$ are provided in table A.8 in the appendix.

The tank pressure is scaled by the characteristic pressure. Considering the reference density $\varrho_{U,\mathrm{ref}}$ provided in table 3.2 in section 3.5, the scaled pressure yields

$$p^* = \frac{p}{\varrho_{U,\mathrm{ref}} \, R_s \, \vartheta_{\mathrm{ref}}} \tag{4.11}$$

where R_s is the specific gas constant provided in table 3.3. The time is scaled by the reciprocal of the excitation frequency as defined in equation (2.71) representing the main driving force for the sloshing. Thus,

$$t^* = \frac{t}{\tau} \tag{4.12}$$

where $\tau = 1/f = 0.714$ s is the characteristic time. The start of the excitation corresponds to the point of time when $t^* = 0$. The excitation of the system is assumed to be a measure of the impact on the expected pressure drop in the tank. This might be modified considering other sloshing modes including the swirl mode.

In the temperature distributions presented here, the height $z/R = 0$ corresponds to the tank bottom, while the free liquid surface is always located at $z/R = 2$. Furthermore, $z/R = 4.48$ gives the position of the tank lid and therefore the total height of the tank. Lists for the reference quantities are provided in table A.8, A.9 and A.10 in the appendix.

4.3.2 Self-pressurization

Self-pressurization in cryogenic liquid tanks mostly occurs when heat from the environment is entering the closed system (either by conduction, convection or radiation). Of course, this effect is stronger for less effective tank insulation. Depending on the latent heat of vaporization Δh_v, incoming heat provokes the evaporation of liquid at the free surface leading to the rise of the tank pressure. This case may accord to the ground and ascent phase where the upper stage propellant tanks are fully loaded containing only liquid and its vapor. The tank pressure is mainly influenced by the thermal heat fluxes flowing through the tank walls.

An illustration of the self-pressurized system in laboratory size is provided in figure 4.4 showing the dewar tank including its instrumentation. After adjusting the required fill level with liquid nitrogen (LN_2) from the storage reservoir, the tank lid is closed, so that the tank pressure can increase. As introduced in chapter 3, it is assumed that the heat flows into the tank through the polyoxymethylene (POM) tank lid and conducts through the inner glass wall to the liquid surface. The total heat flow that enters the tank is measured to be approximately

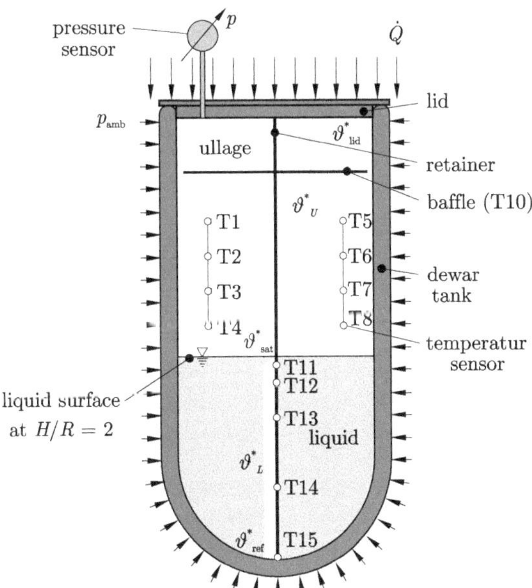

Figure 4.4: Principle for self-pressurizing the dewar tank including the instrumentation in the liquid and in the ullage. The incoming heat flux is indicated by the arrows on the gray area outside the tank. The locations of the temperature sensors are marked by the open dots in the liquid and in the ullage. Qualitative temperature distributions in the liquid and in the ullage are indicated by light gray areas inside the tank.

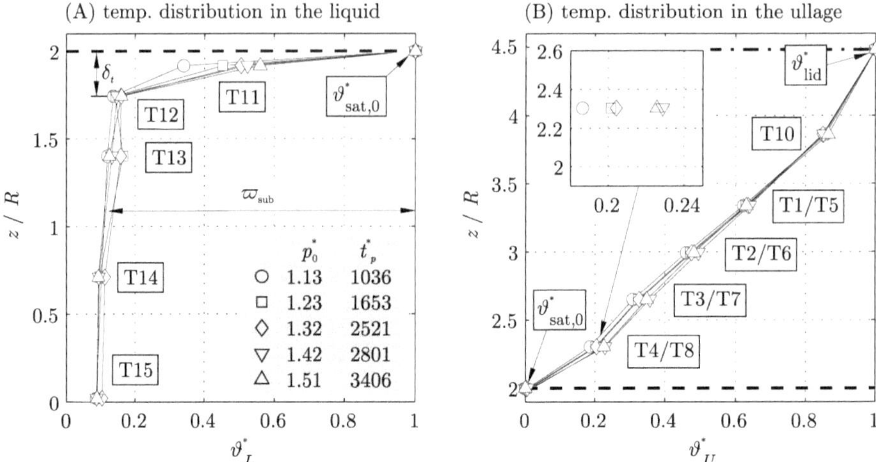

Figure 4.5: Temperature distribution after self-pressurization reaching the required initial pressure p_0^* in the liquid (A) and in the ullage (B). The dashed lines indicate the position of the free liquid surface, while the dash-dot line indicate the position of the tank lid. Open symbols correspond to the temperature data indicating different initial pressure levels. The solid lines merely emphasize the distribution to enhance the readability. The inset in (B) highlights the pressure depended temperature distribution at sensor T4/T8.

$\dot{Q}_{\text{heat}} = 5.9$ W. Regarding this $\dot{Q}$, it takes between 12 and 45 minutes to pressurize the tank depending on the according initial pressure p_0. Further information about the duration of the pressurization t_p^* is provided in the legend of figure 4.5 (A) and in table A.2 in the appendix.

According to the increasing tank pressure, the saturation temperature at the liquid surface increases as well. The liquid temperature at the liquid surface is directly coupled to the actual tank pressure. The relation is described by the CLAUSIUS CLAPEYRON law that is introduced in equation (2.82). Furthermore, the temperature of the liquid bulk remains approximately constant.

The temperature distributions in the tank after self-pressurization for the corresponding dimensionless initial pressure p_0^* are shown in figure 4.5. Here, the temperatures that are measured at different sensor positions T1 to T8 (T10) in the ullage and T11 to T15 in the liquid are plotted along the scaled tank height z/R.

The temperature distribution observed in the liquid is shown in figure 4.5 (A). The symbols here correspond to different initial pressures p_0^*. However, while the temperature at the liquid surface is ϑ_{sat}, the layers below are colder showing a higher temperature gradient between the liquid surface and the liquid bulk. This region is defined as thermal boundary layer with depth δ_t. For self-pressurization, the boundary layer depth observed here is in the order of

4 cm depending on the initial pressure. In deeper regions at sensor positions T12 and T13, the gradient decreases, while the temperature in the bulk at T14 and T15 does not change significantly. The difference between the temperature at the surface and the bulk temperature is defined as degree of subcooling that is for self-pressurization $\varpi_{\mathrm{sub}} \approx 0.9$.

The temperature distribution in the ullage is plotted in figure 4.5 (B) along the scaled tank height z/R. The temperature $\vartheta_U^* = 0$ on the x-axis corresponds to $\vartheta_U = \vartheta_{\mathrm{ref}}$, while $\vartheta_U^* = 1$ corresponds to the inner tank lid temperature that is $\vartheta_{\mathrm{lid}} \approx 280$ K. On the tank height scale, $z/R = 2$ corresponds to the free liquid surface, while $z/R = 4.48$ gives the position of the tank lid and therefore the total height of the tank.

A clear linear temperature stratification is observed between the temperature sensors T4/T8 and the sensor at the lid T9. The temperature gradient appears slightly lower in the region between the baffle and the tank lid. This can be explained by the fact that the baffle represents an obstacle in the upper ullage region. Except of the narrow gap between the baffle and the wall (2 – 3 mm), this region is shielded from the cooler vapor below. Experiments performed with a heated baffle at the same location show that temperature variations in the upper ullage strata do not influence the heat and mass transfer in the liquid/vapor transition at the free surface on time scales, which are considered here. Supplementary, at the temperature couple T4/T8 that is as closest to the liquid surface, the ullage temperature significantly depends on the corresponding initial pressure p_0^*, as shown in the inset of figure 4.5 (B). Between the free liquid surface at $z/R = 2$ and T4/T8, the temperature gradient is slightly higher. Here, the evaporating liquid has a significant impact on the vapor above the free surface. This confirms previous results observed by BARNETT [14], KUMAR et al. [38] and LACAPERE et al. [39]. Furthermore, at the free liquid surface the vapor has saturation temperature, so that $\vartheta_U = \vartheta_L = \vartheta_{\mathrm{sat}}$.

The dimensionless pressure development for a self-pressurized system is shown in figure 4.6 providing the scaled tank pressure p^* plotted versus scaled time t^*. Here, the development provided in figure 4.6 (A) shows the pressurization phase of the system where the liquid surface is undisturbed. After closing the tank under normal pressure conditions indicated by the dash-dot line, the pressure increases linearly until reaching the corresponding initial pressure p_0^* at $t^* = 0$. At this point the excitation is initiated and the liquid starts to slosh. The pressure development during the sloshing phase is shown in figure 4.6 (B). As a matter of fact, the first surface wave amplitudes are stronger before the oscillation settles down to reach the steady state after approximately $t^* \approx 250$. After starting the excitation in the self-pressurized system, it is observed that firstly the pressure slightly increases as shown for $t^* < 10$ in figure 4.6 (B). This can be traced back to the fact that the warmer tank wall upside of the undisturbed liquid surface level provokes the evaporation of the liquid film left by the sloshing liquid moving downwards. This continues for a few oscillations until the entire heat from the tank wall in this region is distributed. Then, the evaporation effect decays and instead the vapor condensation caused by the temperature decrease of the surface temperature dominates the pressure development in

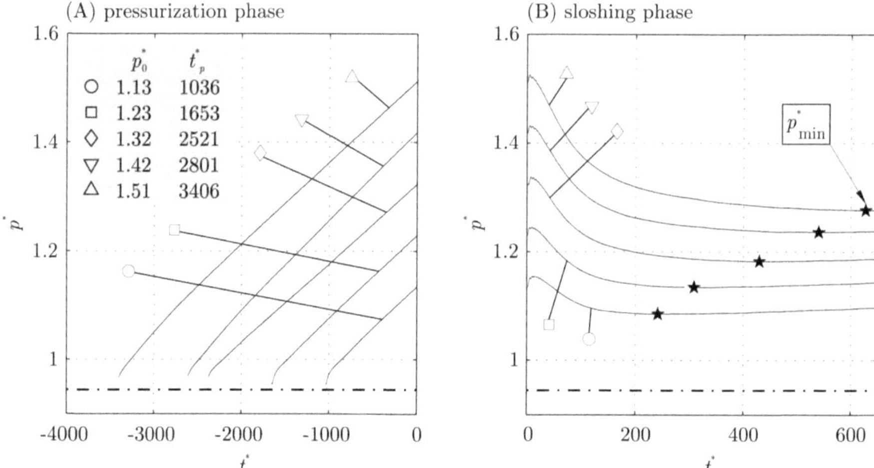

Figure 4.6: (A) Pressure development during the self-pressurization phase. (B) Pressure development during the sloshing phase. The solid lines correspond to the experimental pressure data, while the open symbols merely indicate the according initial pressure levels p_0^*. The ★ symbols indicate the pressure minima p_{min}^*. The dash-dot lines correspond to the norm pressure level.

the tank. The tank pressure decreases until the pressure minimum is reached. At this point, the thermal conditions in the liquid surface layer are re-balanced. Hence, the pressure decrease stops. This is indicated by ★ symbols. After reaching the pressure minimum p_{min}^* the tank pressure re-increases due to the external heat entering the tank.

In accordance to the pressure development provided in figure 4.6, the liquid temperature development in the vicinity of the free surface is plotted versus scaled time in figure 4.7. The pressurization phase is shown in figure 4.7 (A) for $t^* < 0$. The liquid temperatures correspond to the temperatures measured at sensor position T11, while the thick solid lines indicate the saturation temperature of the free surface. The saturation temperature at the free surface is determined from the tank pressure by applying the CLAUSIUS CLAPEYRON relation that is introduced in equation (2.82). Again here, $\vartheta_L^* = 1$ corresponds to the saturation temperature $\vartheta_{sat,0}$ at the liquid surface for the initial pressure p_0, while $\vartheta_L^* = 0$ accords to the reference temperature ϑ_{ref}. While the temperature at the free surface rises in accordance to the tank pressure, the liquid temperature below at T11 rises more slowly due to thermal inertia.

The excitation is started at $t^* = 0$ reaching the corresponding initial pressure. The liquid temperature development in the vicinity of the free surface during the sloshing phase is shown in figure 4.7 (B). Here, the saturation temperature at the free liquid surface is indicated by the thick solid lines corresponding to the actual tank pressure. With the initiation of liquid

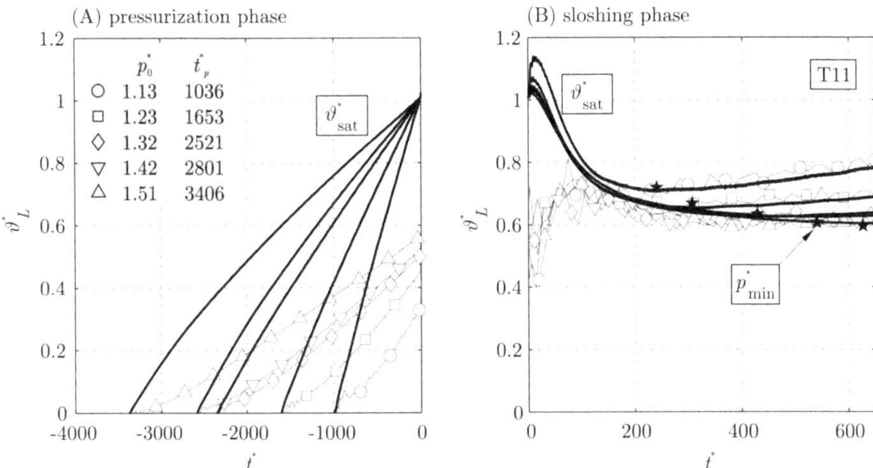

Figure 4.7: Temperature development in the liquid at sensor position T11 during self-pressurization experiments; during the pressurization phase (A) and during the following sloshing phase (B). The thin solid lines correspond to the experimental temperature data, while the open symbols shall merely enhance the readability of the temperature development for the according initial pressure levels p^*_0. The ★ symbols correspond to the pressure minima p^*_{min}. The thick solid lines correspond to the saturation temperature at the the liquid surface determined from the tank pressure development according to equation (2.82).

sloshing, the temperature at T11 rapidly increases as a result of mixing effects with warmer liquid from the surface. On the other side, the temperature of the liquid at the free surface and thus the tank pressure decreases. After a while, the pressure minimum is reached where the temperatures at T11 and at the free surface are approximately equaled leading to an almost homogeneously developed thermal stratification in this region. Reaching this point includes that the thermal equilibrium in the liquid/vapor interface is quasi re-established. From here, vaporization effects caused by the heat coming through the tank lid and tank walls gain in impact leading to the re-increase of the tank pressure and therefore the liquid temperature in the thermal boundary layer in the vicinity of the free liquid surface.

The corresponding temperature development in the ullage is provided in figure 4.8 showing the averaged temperature for the sensor couples T1/T5 to T4/T8. During the pressurization phase, the ullage temperature slowly increases according to the actual tank pressure as shown in figure 4.8 (A). The symbols correspond to the according initial pressure values and therefore to different pressurization times t^*_p. Again here, the y-axis is scaled, so that $\vartheta^*_U = 0$ corresponds to the saturation temperature of the according initial pressure $\vartheta_{\mathrm{sat},0}$, while $\vartheta^*_U = 1$ corresponds to the inner tank lid temperature $\vartheta^*_{\mathrm{lid}}$. While there is not much change in the upper strata, the temperatures at T3/T7 and T4/T8 show a small dependency on the initial pressure. Here,

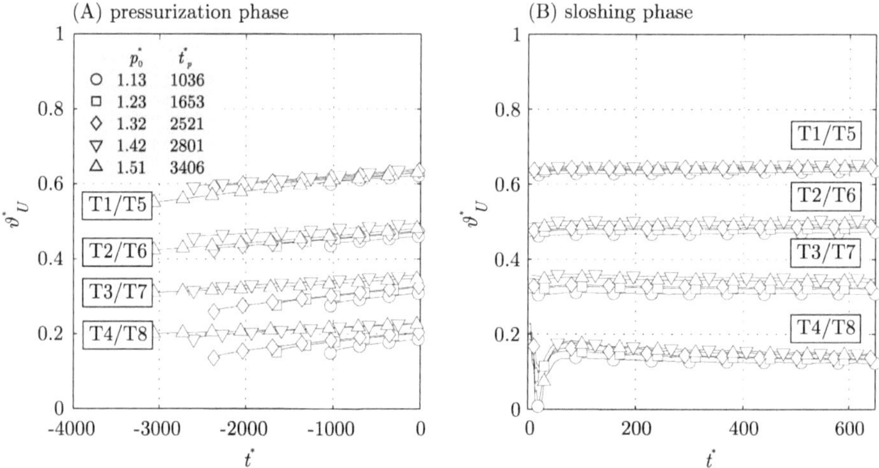

Figure 4.8: Temperature development in the ullage for self-pressurization experiments; during the pressurization phase (A) and during the following sloshing phase (B). The ullage temperatures are averaged for the respective sensor pairings. The solid lines correspond to the experimental temperature data, while the open symbols shall merely enhance the readability of the temperature development for the according initial pressure levels. Sensor contact to the liquid is indicated by $\vartheta_U^* < 0.1$.

the ullage temperature is smaller for the smaller p_0^*. However, at the end of the pressurization phase reaching $t^* = 0$, the ullage is clearly stratified with the lower temperature in the vicinity of the free surface and the higher temperature in the upper strata.

After starting the excitation, which is shown in figure 4.8 (B), only the lower regions at T4/T8 are affected by the sloshing liquid. Here, the ullage temperature decreases due to the cooling-down of the liquid below, so that the ullage cools down as well, while the tank pressure decreases. Temperatures $\vartheta_U^* < 0.1$ indicate that the temperature sensor is temporarily in contact with the sloshing liquid. In the upper regions on level of T3/T7 to T1/T5, the sloshing liquid does not have a significant influence on the temperature development in the ullage showing approximately constant values.

Back to figure 4.6 (B) and figure 4.7 (B), the ★ symbols indicate the point when the local pressure minimum $p_{\min}^*$ is reached. From here, the pressure re-increases due to the heat flowing through the tank walls forcing evaporation although the liquid is still in motion. Thus, evaporation effects dominate over condensation, so that the tank pressure must rise. The temperature distribution in the liquid at this point is shown in figure 4.9 (A). In comparison to the liquid temperature distribution at $t^* = 0$ provided in figure 4.5 (A), the liquid temperature here, particularly in the vicinity of the free surface at T11 and T12, is significantly higher than

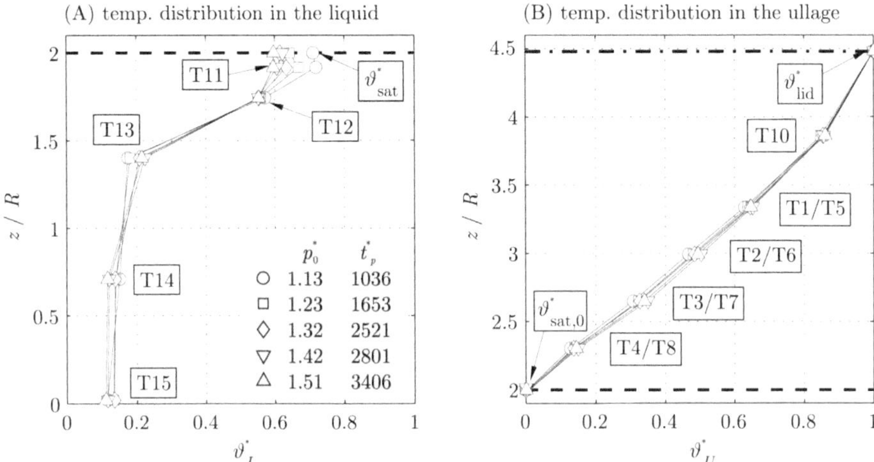

Figure 4.9: Temperature distribution for a self-pressurized tank reaching the minimum pressure $p^* = p^*_{\min}$ in the liquid (A) and in the ullage (B). The dashed lines indicate the position of the free liquid surface, while the dash-dot line indicate the position of the tank lid. Open symbols correspond to the temperature data indicating different initial pressure levels. The solid lines merely emphasize the distribution to enhance the readability.

observed in the initial state at $t^* = 0$. The temperature rise here is the consequence of the mixing with warmer liquid from above, which is supported by the latent heat released by the condensation. Furthermore, a more homogeneous stratification is observed in this region. This confirms previous studies by LACAPERE et al. [39] and DAS & HOPFINGER [26].

The temperature distribution in the ullage observed at the pressure minima is provided in figure 4.9 (B). Except for the lower strata below T4/T8 where the ullage temperature is slightly lower than observed for $t^* = 0$ in figure 4.5 (B), the temperature distribution in the ullage here remains stable without significant variations under the impact of sloshing. This may be different by considering other sloshing modes such as the swirling mode producing more chaotic liquid motion.

On the next pages, the thermodynamic balance of the self-pressurized system is provided. The dashed lines in figure 4.10 represent the control volume boundary to balance the internal (liquid and ullage) and external (heat flow into the tank) energies. The total balance of energy during the self-pressurization experiments is composed of two parts, the energy balance for the liquid and the energy balance for the ullage as shown in figure 4.10 yielding

$$dQ_{\mathrm{tot}} = dQ_L + dQ_U \, . \tag{4.13}$$

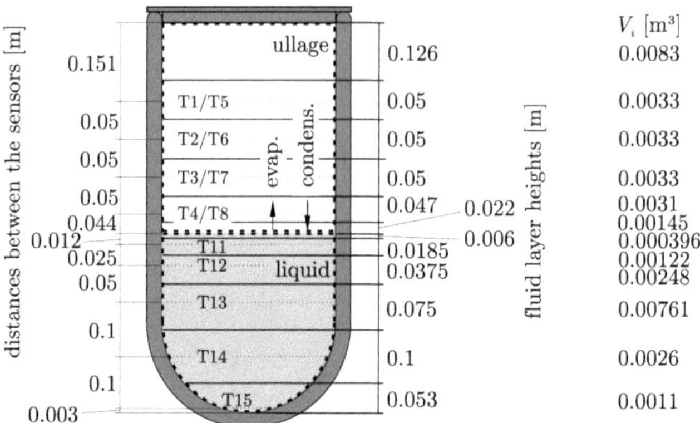

Figure 4.10: Control volumes inside of the tank for self-pressurization in the liquid and in the ullage. The measures on the left side correspond to the distances between the temperature sensor, while the measures on the right side correspond to the fluid layer heights. The column on the far right side includes the according volume for each layer.

The liquid can be considered as open system. Based on equation (2.14), the balance for the liquid can be rewritten in general by

$$dQ_L = dU_{e,L} + \sum m_{L,\text{out}} \, h_{e,L,\text{out}} - \sum m_{L,\text{in}} \, h_{e,L,\text{in}} \qquad (4.14)$$

where mass can either enter (condensation) or leave (evaporation) the control volume via the free surface. As introduced in equation (2.17), the inner energy of the liquid can be expressed by the enthalpy, so that

$$dU_{e,L} = dH_{e,L} - p \, dV_L - V_L \, dp \qquad (4.15)$$

where $p \, dV_L = 0$ for constant volumes. Thus, the expression for the inner energy in the liquid simplifies

$$dU_{e,L} = dH_{e,L} - V_L \, dp \,. \qquad (4.16)$$

The change of the enthalpy is defined as difference between the enthalpy in state (II), e.g. end of pressurization, and state (I), e.g. start of pressurization, so that

$$dH_{e,L} = H_{e,L}^{(II)} - H_{e,L}^{(I)} = m_L^{(II)} \, h_{e,L}^{(II)} - m_L^{(I)} \, h_{e,L}^{(I)} \qquad (4.17)$$

with

$$m_L = \sum_k \varrho_{L,k} \left(\vartheta_{L,k} \right) V_{L,k} \qquad (4.18)$$

defined as the mass of the liquid. As shown in figure 4.10, the liquid is subdivided into k layers based on the distances between the temperature sensors. Referring to the fluid database [41], the specific enthalpy $h_{e,L}(\vartheta_L)$ for each layer is determined from the measured temperatures at sensor position T11 to T15. The liquid mass in each layer is determined by the liquid volume based on the fluid layer heights as shown in figure 4.10 and the temperature depended liquid density $\varrho_L(\vartheta_L)$ taken from [41] as well. Thus, the enthalpy for the entire liquid volume reads

$$dH_{e,L} = \sum_k m_{L,k}^{(II)} h_{e,L,k}^{(II)} - \sum_k m_{L,k}^{(I)} h_{e,L,k}^{(I)} . \tag{4.19}$$

In the following, the energy balance for the liquid is defined in detail for the different phases; pressurization and sloshing. For pressurization, the energy balance for the liquid gives

$$dQ_{L,\text{press}} = dH_{e,L} - V_L\, dp + m_{\text{evap}} \left[h_{e,L,\text{evap}} + \Delta h_v \right] \tag{4.20}$$

where $m_{\text{evap}} \left[h_{e,L,\text{evap}} + \Delta h_v \right]$ is the converted vapor energy that is removed from the liquid due to evaporation during self-pressurization. In this connection, $h_{e,L,\text{evap}}$ is the liquid enthalpy at surface level corresponding to the liquid leaving the liquid phase and Δh_v is the latent heat of vaporization required for the phase change from liquid to vapor. Equation (4.14) assumes that the mass m_{evap} leaves the control volume as liquid. The conversion of that liquid mass into a vapor mass requires the latent heat of evaporation. This is considered in equation (4.20). Please note that this approach is equivalent to

$$dQ_{L,\text{press}} = dH_{e,L} - V_L\, dp + m_{\text{evap}}\, h_{e,U,\text{evap}} . \tag{4.21}$$

The enthalpy of the vapor mass leaving the lower control volume is then added to the upper control volume as shown later in equation (4.29). The latent heat of evaporation must be provided by the lower control volume, i.e. the heat entering the lower control volume increases the internal energy and provided $m_{\text{evap}} \Delta h_v$. Furthermore, the determination of the mass of evaporated nitrogen m_{evap} during self-pressurization is described afterwards in the energy balance for the ullage.

By neglecting evaporation due to the short time scale during sloshing, the energy balance during the sloshing phase reads

$$dQ_{L,\text{slosh}} = dH_{e,L} - V_L\, dp - m_{\text{condens}} \left[h_{e,L,\text{condens}} + \Delta h_v \right] \tag{4.22}$$

where $m_{\text{condens}} \left[h_{e,L,\text{condens}} + \Delta h_v \right]$ is the energy transferred into the liquid by vapor condensation during sloshing. Analogous to the pressurization phase, $h_{e,L,\text{condens}}$ corresponds to the liquid enthalpy at surface level of the condensed mass to be added to the liquid. During the phase change from vapor to liquid, the latent heat Δh_v is released and therefore absorbed by the liquid. Similar to the pressurization phase, this approach is equivalent to

$$dQ_{L,\text{slosh}} = dH_{e,L} - V_L\,dp - m_{\text{condens}}\,h_{e,U,\text{condens}} \qquad (4.23)$$

where the enthalpy of the vapor mass entering the lower control volume is removed from the upper control volume. The released latent heat of evaporation is absorbed by the lower control volume. Furthermore, the determination of the condensed mass is described afterwards in the energy balance for the ullage.

Also the ullage is considered as an open system where vapor can either enter (evaporation) or leave (condensation) the control volume via the free surface. Thus, the energy balance for the ullage is defined in general by

$$dQ_U = dU_{e,U} + \sum m_{U,\text{out}}\,h_{e,U,\text{out}} - \sum m_{U,\text{in}}\,h_{e,U,\text{in}} \qquad (4.24)$$

regarding evaporation during pressurization in the source term $m_{\text{in}}\,h_{e,\text{in}}$ and condensation during sloshing in the sink term $m_{\text{out}}\,h_{e,\text{out}}$. As introduced in equation (2.17), the inner energy of the ullage can be expressed by the enthalpy yielding

$$dU_{e,U} = dH_{e,U} - p\,dV_U - V_U\,dp\,, \qquad (4.25)$$

again, where $p\,dV_U = 0$ for constant volumes. As well in the ullage, the change of the enthalpy is defined as difference between the enthalpy in state (I) and state (II), so that

$$dH_{e,U} = H_{e,U}^{(II)} - H_{e,U}^{(I)} = m_U^{(II)}\,h_{e,U}^{(II)} - m_U^{(I)}\,h_{e,U}^{(I)}\,. \qquad (4.26)$$

The mass of the ullage gas (vapor) m_U is determined by

$$m_U = \sum_k \varrho_{U,k}\,(\vartheta_{U,k})\,V_{U,k}\,. \qquad (4.27)$$

The maximum error of the approximation given in equation (4.27) is about 5% with respect to a linear temperature stratification of the ullage with ϑ_{sat} at the free surface and ϑ_{lid} at the tank lid as shown in figure 4.5 (B).

As shown in figure 4.10, the ullage is also subdivided into k layers based on the distances between the temperature sensors. Again, the specific enthalpy $h_{e,U}(\vartheta_U)$ for each layer is determined from the measured temperature at sensor position T1/T5 to T4/T8 taken from [41]. The gas mass in each layer is determined by the gas volume based on the liquid layer height as shown in figure 4.10 and the temperature depended ullage gas density $\varrho_U(\vartheta_U)$ is taken from [41] as well. Thus, for the entire ullage volume, the enthalpy reads

$$dH_{e,U} = \sum_k m^{(II)}_{U,k} h^{II}_{e,U,k} - \sum_k m^{(I)}_{U,k} h^{(I)}_{e,U,k} \, . \tag{4.28}$$

In the following, the energy balance for the ullage is defined in detail for the different phases; pressurization and sloshing. For pressurization, the energy balance in the ullage reads

$$dQ_{U,\mathrm{press}} = dH_{e,U} - V_U \, dp - m_{\mathrm{evap}} h_{e,U,\mathrm{evap}} \tag{4.29}$$

where $m_{\mathrm{evap}} h_{e,U,\mathrm{evap}}$ is the energy that is added to the ullage by the evaporated liquid during the self-pressurization phase while $h_{e,U,\mathrm{evap}}$ corresponds to the vapor enthalpy at surface level. The mass of the evaporated nitrogen to pressurize the tank is determined by integrating the respective vapor mass over all gas layer (see figure 4.10), so that

$$m_{\mathrm{evap}} = m^{(II)}_U - m^{(I)}_U = \sum_k m^{(II)}_{U,k} - \sum_k m^{(I)}_{U,k} \, , \tag{4.30}$$

while state (I) corresponds to the start of pressurization and state (II) corresponds to the end of pressurization. Again, the vapor mass in each layer is determined by

$$m_{U,k} = \varrho_{U,k} \left(\vartheta_{U,k} \right) V_{U,k} \, . \tag{4.31}$$

By neglecting evaporation due to the short time scale, the energy balance during the sloshing phase reads

$$dQ_{U,\mathrm{slosh}} = dH_{e,U} - V_U \, dp + m_{\mathrm{condens}} h_{e,U,\mathrm{condens}} \tag{4.32}$$

where $m_{\mathrm{condens}} h_{e,U,\mathrm{condens}}$ is the vapor energy released by vapor condensation during sloshing. Analogous to the pressurization phase, $h_{e,U,\mathrm{condens}}$ corresponds to the ullage enthalpy at surface level. Furthermore, the condensed mass is determined by integrating the vapor mass over all layers (see figure 4.10), so that

$$m_{\mathrm{condens}} = m^{(II)}_U - m^{(I)}_U = \sum_k m^{(II)}_{U,k} - \sum_k m^{(I)}_{U,k} \tag{4.33}$$

where state (I) corresponds to the start of sloshing and state (II) corresponds to the point when the minimum pressure is reached. The vapor mass in each layer is determined according to equation (4.31).

The energy balance for the self-pressurized system during the pressurization phase is shown in figure 4.11 (A). During this phase, the total energy difference of the system is equal to the total heat that flows into the tank, so that $dQ_{\mathrm{tot}} = \dot{Q}_{\mathrm{heat}} \, dt$. This is depicted by the $\circ$ symbols corresponding to a heat transfer rate between $\dot{Q}_{\mathrm{heat}} = 9.22 \ldots 10.48$ W (see table 4.4), which is slightly higher than the measured heat transfer rate introduced in chapter 3 that gives

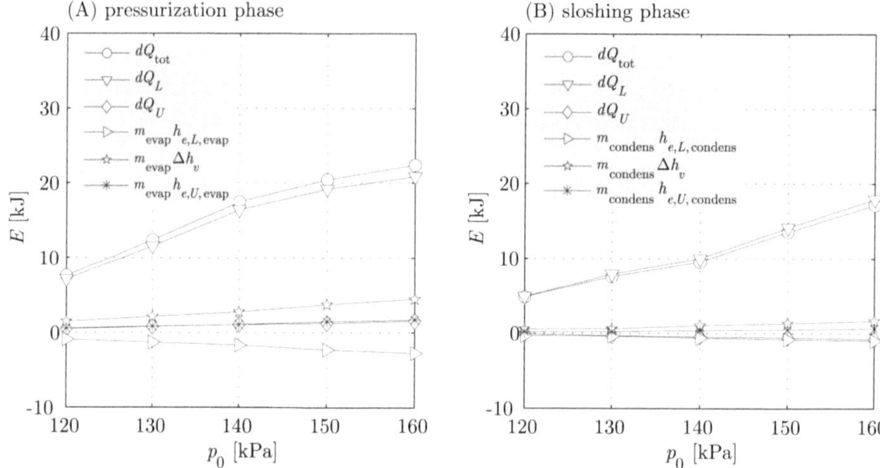

Figure 4.11: Energy balance of self-pressurized systems for the pressurization phase (A) and for the sloshing phase (B). The solid lines connect the data points to enhance the readability.

$\dot{Q}_{\text{heat}} = 5.9$ W and is based on evaporation in an open system only. The energy balance for the liquid introduced in equation (4.16) corresponds to the ∇ symbols, while the energy balance for the ullage introduced in equation (4.29) corresponds to the $\Diamond$ symbols. During pressurization, the larger fraction of energy is absorbed by the liquid having the higher mass fraction. The ullage is already stratified at the beginning of self-pressurization and increases only slightly in terms of temperature and therefore also in terms of energy. The amount of energy that transfers from the liquid to the ullage during self-pressurization by evaporation is $m_{\text{evap}} h_{e,L,\text{evap}}$ in the liquid ($\triangleright$ symbols) and $m_{\text{evap}} h_{e,U,\text{evap}}$ in the ullage ($*$ symbols). The jump between the enthalpy of liquid mass leaving the liquid phase due to evaporation on the one side and the enthalpy of vapor entering the ullage on the other side expresses the phase change and corresponds to the latent heat of evaporation Δh_v, so that

$$\Delta h_v = h_{e,U,\text{evap}} - h_{e,L,\text{evap}} \,. \tag{4.34}$$

The energy balance for the self-pressurized system during the sloshing phase is provided in figure 4.11 (B). Here, the process is driven by condensation initiated by the mixing of the liquid in the thermal boundary layer below the free surface. Consequently, the surface temperature and therefore the saturation temperature decrease. Due to mixing and absorbing the latent heat, the averaged temperature of the liquid slightly increases leading to an increase of the energy in the liquid indicated by the ∇ symbols. This is different in the ullage where the energy difference is indicated by the $\Diamond$ symbols. Led by condensation and the ullage temperature cooling-down, the energy in the ullage must decrease as well. Values for the pressurization

Table 4.4: Heat input and phase change energy during self-pressurization experiments.

		120	130	140	150	160
	p_0 [kPa]	120	130	140	150	160
	dQ_{tot} [kJ]	7.62	12.36	17.46	20.44	22.42
	t_p [s]	740	1180	1800	2000	2432
	$\dot{Q}_{\text{heat}}$ [W]	10.3	10.48	9.7	10.22	9.22
press	$m_{\text{evap}}\, h_{e,L,\text{evap}}$ [kJ]	-0.91	-1.31	-1.68	-2.3	-2.74
	$m_{\text{evap}}\, \Delta h_v$ [kJ]	1.49	2.14	2.75	3.76	4.48
	$m_{\text{evap}}\, h_{e,U,\text{evap}}$ [kJ]	0.58	0.83	1.07	1.46	1.74
	m_{evap} [g]	7.47	10.75	13.82	18.89	22.48
	Δh_v [kJ kg^{-1}]	199.2	199.19	199.04	199.04	199.07
slosh	$m_{\text{condens}}\, h_{e,L,\text{condens}}$ [kJ]	-0.35	-0.38	-0.63	-0.81	-1.01
	$m_{\text{condens}}\, \Delta h_v$ [kJ]	0.57	0.62	1.04	1.34	1.66
	$m_{\text{condens}}\, h_{e,U,\text{condens}}$ [kJ]	0.22	0.24	0.41	0.53	0.65
	m_{condens} [g]	2.85	3.09	5.20	6.66	8.29
	Δh_v [kJ kg^{-1}]	200.02	200.29	200.33	200.6	200.86

phase and the sloshing phase are presented in table 4.4 as well as in table A.19 and table A.20 in the appendix.

Obviously, the pressure drop that occurs under the impact of sloshing is superimposed by two different effects; the ullage cooling-down as well as condensation effects caused by the decrease of the surface temperature as shown in figure 4.12. The physical impact of the pressure drop is directly measured by the pressure sensor in the tank lid. Thus, the total change of the tank pressure is given by $dp_{\text{press drop}}$ for the given ullage volume as indicated by the data corresponding to the full ◄ symbols. By knowing the mean ullage temperature and the mean mass distribution from the ullage gas densities, the pressure loss fraction caused by the temperature decrease of the ullage gas is determined by means of the ideal gas law, so that

$$dp_{\text{ullage temp}} = \frac{1}{V_U}\, m_U\, R_s\, d\bar{\vartheta}_U \tag{4.35}$$

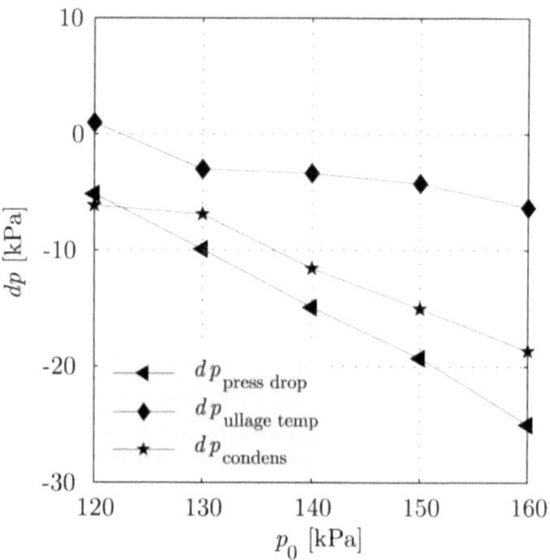

Figure 4.12: Composition of the pressure drop for the self-pressurized systems showing the pressure change in the ullage. The solid lines connect the data points to increase the readability.

where R_s is the specific gas constant for GN_2 provided in table 3.3. The mean temperature of the ullage is determined by

$$\bar{\vartheta}_U = \frac{1}{m_U} \sum_k \vartheta_{U,k}\, \varrho_{U,k}\, V_{U,k} \qquad (4.36)$$

where the ullage mass m_U is determined by equation (4.27). The data for $dp_{\text{ullage temp}}$ is indicated by the full ♦ symbols. Except for the smallest initial pressure where the ullage temperature slightly increases, the energy difference due to the ullage temperature cooling-down is approximately independent from the initial pressure and only depends on the sloshing. Thus, the energy dissipated by the ullage due to condensation is

$$dp_{\text{condens}} = dp_{\text{press drop}} - dp_{\text{ullage temp}} \qquad (4.37)$$

indicated by the full ★ symbols. The fraction of condensation is about 75% of the total pressure loss, while the fraction of ullage temperature decrease is about 25%.

4.3.3 External Nitrogen Pressurization

The external nitrogen pressurization of a tank filled with liquid nitrogen is in analogy to a spacecraft using a one-species system where liquid propellant is externally vaporized to pressurize the propellant tank. The vaporized propellant may be induced into the tank from gas generators fed by hot parts of the rocket engine. This scenario may be conceivable during the upper stage boost phase where the drained volume in the hydrogen tank compartment is compensated by vaporized LH_2 that is utilized as fuel in the full size application. The laboratory size experiments that are conducted for this work use liquid nitrogen (LN_2) as substitute for the upper stage propellant. The external pressurization is realized by connecting a common high pressure nitrogen gas bottle to the test tank. For safety issue, a pressure reducer is used to limit the pressure of the gas bottle to $p_{\mathrm{max}} = 200$ kPa. The pressurization mass flow rate corresponds to $\dot{m}_{\mathrm{GN_2}} = 0.105$ g s^{-1}, which is controlled by a flow meter at the gas inlet on the tank lid.

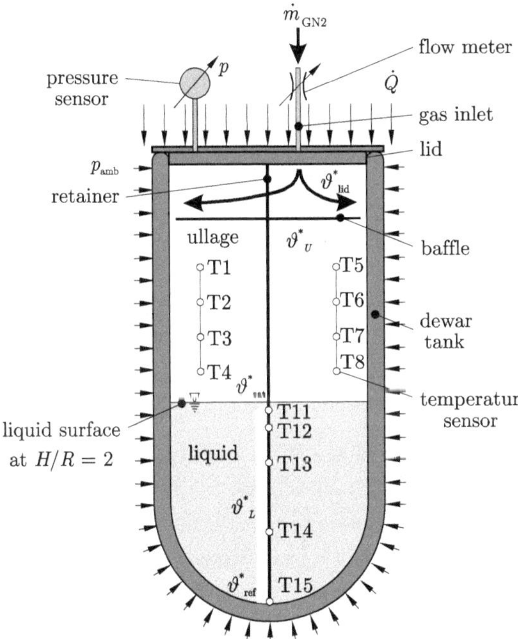

Figure 4.13: Principle for pressurizing the tank with GN_2 including the instrumentation in the liquid and in the ullage. The heat flowing into the tank is indicated by the gray area outside the tank and by small arrows. The locations of the temperature sensors are marked by the open dots in the liquid and in the ullage. Qualitative temperature distributions are indicated by light gray areas inside the tank. The GN_2 is fed in the tank through the gas inlet integrated in the lid; the flow rate of $\dot{m}_{\mathrm{GN_2}} = 0.105$ g s^{-1} is controlled by a flow meter.

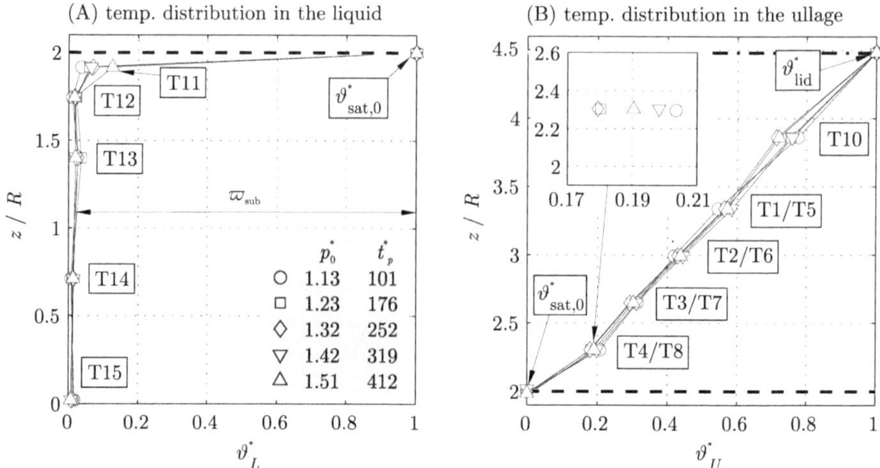

Figure 4.14: Temperature distribution after GN_2 pressurization reaching the required initial pressure p^*_0 in the liquid (A) and in the ullage (B). The dashed lines indicate the position of the free liquid surface, while the dash-dot line indicate the position of the tank lid. Open symbols correspond to the temperature data indicating different initial pressure levels. The solid lines merely emphasize the distribution to enhance the readability. The inset in (B) highlights the pressure dependent temperature distribution at sensor T4/T8.

The temperature of the pressurization gas can be considered as constant at $\vartheta_{GN2} \approx 293$ K. An illustration of the external GN_2 pressurized configuration is provided in figure 4.13. Here, the gas is injected through the gas inlet into the tank. To prevent the gas jet from directly hitting the liquid surface, the injected gas is conveyed on the baffle to be equally distributed along the tank perimeter. In this case, the thermal stratification in the ullage can be conserved to ensure repeatable initial conditions inside the tank.

The experimental procedure starts by filling the pre-cooled tank up to the required tank level, which is defined by $H/R = 2$. Then, the tank lid is closed in order to allow the tank pressure to increase. Using external GN_2 pressurization reduces the duration to reach the required initial pressure by a factor of 10 compared to the self-pressurization case that is driven by external heat entering the tank under similar conditions. The durations for the external GN_2 pressurization are listed in table A.2 in the appendix. During the experimental run, the tank pressure as well as the temperatures in the ullage and in the liquid are logged. In figure 4.13, the temperature sensors in the liquid are indicated by T11 – T15, while the temperature sensors in the ullage are indicated by T1 – T8. The tank pressure p^* is measured by using a pressure sensor located at the tank lid.

The temperature distribution along the scaled tank height z/R after GN_2 pressurization and before starting the excitation is shown in figure 4.14. According to equation (4.9), the scaled liquid temperature distribution ϑ_L^* for the initial condition at $t^* = 0$ is provided in figure 4.14 (A). Contrary to the self-pressurization experiments presented in the previous section, the temperatures below the free liquid surface between sensor positions T11 and T15 are lower. The effect is particularly obvious at sensor position T11 where the liquid temperature varies only slightly from the bulk temperature for small p_0^*. Due to thermal inertia, the development of the thermal boundary layer δ_t below the liquid surface depends on the duration of pressurization and on the initial pressure. As well as in the previous case, it is assumed that the liquid at the free surface must have saturation temperature ϑ_{sat}. For a given tank pressure, the saturation temperature can be determined by the CLAUSIUS CLAPEYRON law introduced in equation (2.82). Due to the applied scaling, the liquid temperature ϑ_L^* is related to the saturation temperature for the corresponding initial pressure. Again, the difference between the temperature at the surface and the bulk temperature is defined as degree of subcooling that is for GN_2 pressurization $\varpi_{\mathrm{sub}} \approx 0.98$.

The temperature distribution in the ullage is provided in figure 4.14 (B). Based on the ullage temperature scaling introduced in equation (4.10), the dimensionless temperature ϑ_U^* is plotted along the scaled tank height. The lower dashed line at $z/R = 2$ corresponds to the free liquid surface whereas the upper dashed line at $z/R = 4.48$ corresponds to the position of the tank lid. The temperature distribution after the GN_2 pressurization phase is similar to the distribution observed after self-pressurization in the previous test case. The development of the temperature stratification in the ullage is supported by forced convection (gas injection) and by the heat that conducts through the circumferential tank walls. However, the baffle below the gas inlet preserves the thermal stratification from being disturbed by the external gas feeding. Regarding the large time scales for self-pressurization, this may be additionally supported by conductive heat transport through the vapor for some extent, which is initiated by the external heat flow coming through the warmer tank lid. Rapid gas feeding into the tank certainly provokes an increase of the ullage temperature. Regarding the shorter time scales for external pressurization, it is assumed that the impact of the conductive heat transport is of minor importance for the development of the thermal stratification in the ullage. Therefore, the temperatures in the upper strata appear lower than observed after self-pressurization. Furthermore, the ullage temperature at T4/T8 is not observed to be a function of the corresponding initial pressure as it was the case in the previous section. Instead, the temperatures at T4/T8 as well as at T3/T7 vary only for a small extent and appear more homogeneous as shown in the inset of figure 4.14 (B). Hence, it is assumed that the development of the ullage temperature profile is mainly driven by the forced convection due to external pressurization, while the variation of the duration of pressurization is not sufficient to show a dependency on the respective tank pressure after GN_2 pressurization.

The dimensionless tank pressure development for external GN_2 pressurization is provided in figure 4.15 where the scaled tank pressure p^* is plotted versus scaled time t^*. Again, the

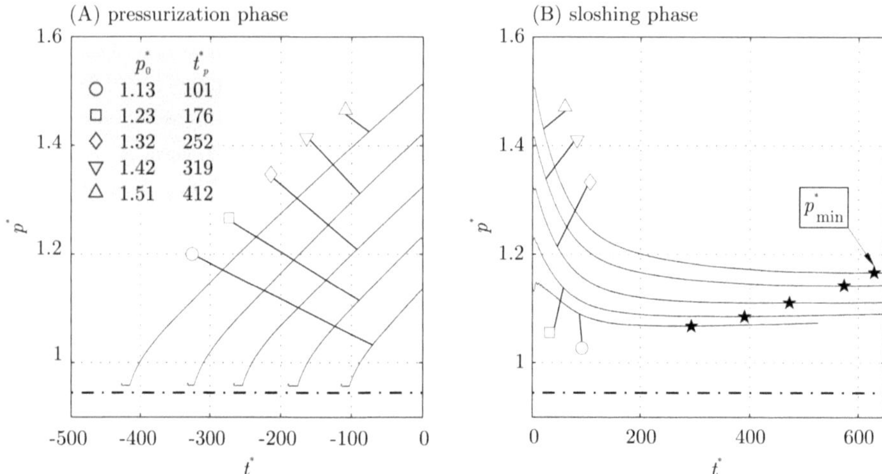

Figure 4.15: (A) Pressure development during the GN_2 pressurization phase. (B) Pressure development during the sloshing phase. The solid lines correspond to the experimental pressure data, while the open symbols only indicate the according initial pressure level. The ★ symbols indicate the pressure minima p_{min}^*. The dash-dot lines correspond to the norm pressure level.

symbols correspond to different initial pressures varying from $p_0^* = 1.13$ to $p_0^* = 1.51$ at $t^* = 0$. The scaled quantities are determined according to equations (4.10) and (4.12). Indicating the pressurization of the tank with GN_2 for $t^* < 0$, this is shown in figure 4.15 (A). Due to the obvious pressure gradients, the time period to reach the desired initial pressure is by a factor 10 smaller than observed during the self-pressurization experiments. This leads to the more subcooled liquid bulk ($\varpi_{sub,\,self-press} \approx 0.9 < \varpi_{sub,\,GN2\,press} \approx 0.98$) with respect to the corresponding initial pressure as shown in figure 4.5 (A) and in figure 4.14 (A). Particularly in the vicinity of the free surface at T11, the scaled liquid temperature ϑ_L^* is by a factor 2 lower than observed after self-pressurization. The system is excited for $t^* > 0$ representing the sloshing phase. The pressure developments under the impact of liquid sloshing are provided in figure 4.15 (B). For the smaller initial pressures, the tank pressure increases slightly during the first oscillations indicated by the small pressure peak at the very beginning of the sequence. As explained in the previous section, this pressure raise can be traced back to evaporation effects, while the liquid gets into contact with the warmer tank walls during the first oscillations. This is different for higher initial pressures. Here, the pressure decreases immediately after starting the excitation. Contrary to the self-pressurized system, which is thermodynamically balanced due to the long duration pressurization time, the duration to pressurize the GN_2 system is much shorter. Nevertheless, the liquid at the free surface is at saturation temperature ϑ_{sat}^*. This is not the case in regions below the free surface as shown in figure 4.14 (A) where the temperatures between T11 and T15 appear constant and subcooled with respect to $\vartheta^*_{\text{sat},0}$, the saturation temperature at $t^* = 0$. The temperature gradient in the thermal boundary layer is higher in GN_2 pressurization systems. However, GN_2 pressurized experiments without sloshing show that the tank pressure decreases slightly after stopping the pressurization [4]. This is caused by the re-establishment of the thermodynamic equilibrium in the thermal boundary layer under the free liquid surface. It is assumed that this decrease at least compensates the vaporization for $p^*_0 > 1.32$ that occurs during the first oscillations when the cold liquid is in contact with the warmer tank wall above the undisturbed liquid surface.

The characteristic pressure drop for $t^* > 0$ is similar to the effect observed for a selfpressurized system whereas the magnitude of the pressure decrease is larger in the GN_2 pressurized case particularly for higher initial pressures. For small initial pressures the development is similar in both cases. The tank pressure stops to decrease when the local pressure minimum p^*_{min} is reached. This is indicated by full asterisk symbols shown in figure 4.15 (B). Here, the liquid in the thermal boundary layer below the free liquid surface at T11 and T12 is approximately homogeneous mixed with the liquid from the surface. Thus, the condensation stops and the tank pressure can re-increase. The pressure gradients dp^*/dt^* [7, 26] right after starting the excitation for $0 \leq t^* \leq 50$ are higher than observed during self-pressurization. For increasing initial pressure the magnitude of the pressure gradient increases as well.

The liquid temperature development in the vicinity of the free liquid surface is provided in figure 4.16. The shown liquid temperatures correspond to the temperatures measured at T11, while the solid thick lines indicate the saturation temperature of the free surface. Those are determined from the tank pressure by applying the CLAUSIUS CLAPEYRON relation introduced in equation (2.82). Again here, $\vartheta^*_L = 1$ corresponds to $\vartheta_{\text{sat},0}$, the saturation temperature at p_0, while $\vartheta^*_L = 0$ corresponds to the reference temperature ϑ_{ref}. However, the pressurization phase is shown in figure 4.16 (A) for $t^* < 0$. While the temperature at the free liquid surface rises in accordance to the actual tank pressure, the liquid temperature below the free surface at T11 rises much slower due to thermal inertia. Therefore, the thermal boundary layer δ_t here is smaller for some extent than in the self-pressurized system where the temperature gradient between the liquid surface and T11 is higher.

The excitation of the tank is started at $t^* = 0$ when the corresponding initial pressure p^*_0 is reached. The temperature development in the liquid at T11 during the sloshing phase is shown in figure 4.16 (B). Again, the temperature at the free liquid surface is indicated by the thick solid lines corresponding to the actual tank pressure. With the initiation of liquid sloshing, the temperature at T11 rapidly increases as a result of the mixing with warmer liquid from the surface. From this mixing, the temperature at the free liquid surface decreases provoking the condensation of vapor in the ullage and the tank pressure decreases as well in accordance to the CLAUSIUS CLAPEYRON relation in equation (2.82). It may be assumed that the released latent heat is dissipated through the liquid explaining the increase of the liquid temperature at T11 for

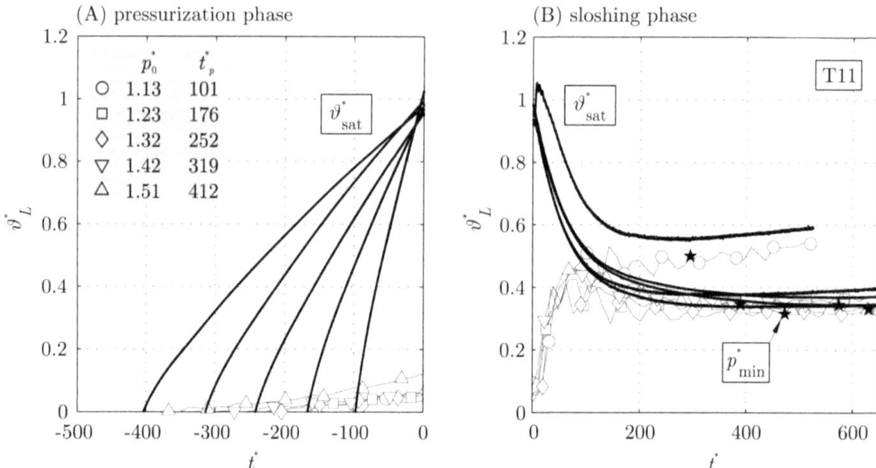

Figure 4.16: Temperature development in the liquid at sensor position T11 during GN_2 pressurization experiments; during the pressurization phase (A) and during the following sloshing phase (B). The thin solid lines correspond to the experimental pressure data, while the open symbols merely indicate the according initial pressure value. The ★ symbols indicate the pressure minima p^*_{min}. The thick solid lines correspond to the saturation temperature at the the liquid surface determined from the tank pressure development according to equation (2.82).

some extent. The local pressure minimum is reached when the tank pressure does not further decrease. This is indicated by ★ symbols in figure 4.16 (B). For the pressure minima, the temperatures at T11 and at the free surface are almost equaled leading to a more homogeneous stratification in this region. Here, the thermal equilibrium in the liquid/vapor transition is re-established. The tank pressure and therefore the liquid temperature in the boundary layer below the free liquid surface can increase again.

The temperature development in the ullage is provided in figure 4.17 showing the averaged temperatures for the sensor couples T1/T5 to T4/T8. The ullage temperature development during the pressurization phase for $t^* < 0$ is shown in figure 4.17 (A). The symbols correspond to different initial pressures p^*_0 and therefore to different pressurization times t^*_p. Again here, the y-axis is scaled, so that $\vartheta^*_U = 0$ corresponds to the saturation temperature of the according initial pressure $\vartheta_{sat,0}$, while $\vartheta^*_U = 1$ corresponds to the inner tank lid temperature ϑ^*_{lid}. As well as observed at the beginning of the self-pressurization experiments, the formation of the temperature stratification occurs already during the filling process of the tank. When the data logging is started, the linear stratification in the ullage is already fully developed. Further temperature increase during GN_2 pressurization can be traced back to the pressure increase in the tank regarding GN_2 as an ideal gas. However, at the end of the pressurization phase

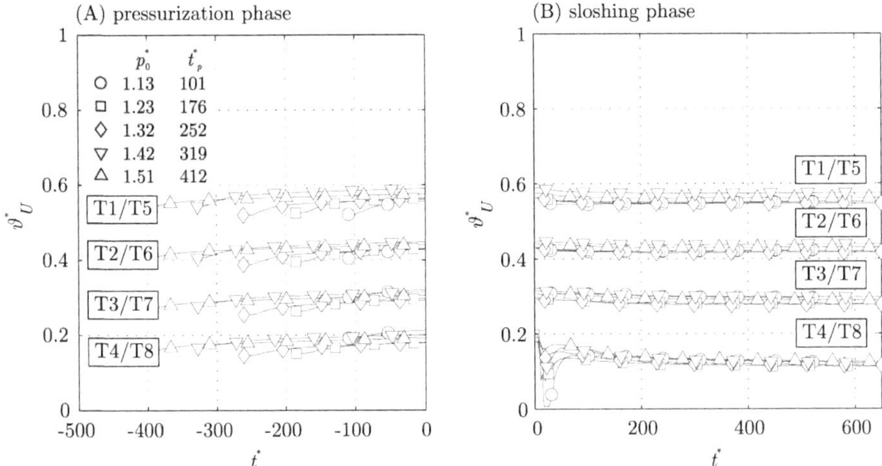

Figure 4.17: Temperature development in the ullage for GN_2 pressurization experiments during the pressurization phase (A) and during the following sloshing phase (B). The symbols indicate different initial pressure levels, while the solid lines correspond to the scaled measurement data. The ullage temperatures are averaged for the corresponding sensor pairings. Sensor contact to the liquid is indicated by $\vartheta_U^* < 0.1$.

at $t^* = 0$, the ullage is clearly stratified showing similar temperatures for all considered initial pressures.

After starting the excitation, which is shown in figure 4.17 (B), only the thermal layers in the lower regions at T4/T8 are affected by the sloshing liquid. Here, the ullage temperature decreases due to heat exchange to the sloshing liquid, so that the ullage cools down, while the tank pressure decreases. Temperature values below $\vartheta_U^* < 0.1$ indicate that the temperature sensor is partly in contact with the sloshing liquid. In the upper regions on level of T3/T7 to T1/T5, the sloshing liquid does not have any significant influences on the temperature development in the ullage. Here, the ullage temperature remains constant.

Back to figure 4.15 (B) and figure 4.16 (B), the ★ symbols indicate the point when the local pressure minimum $p_{\min}^*$ is reached. From this point, the tank pressure re-increases indicating that the pressure development is no longer dominated by condensation effects. Instead, the external heat input gains in influence provoking evaporation at the free liquid surface. The temperature distributions in the liquid for $p_{\min}^*$ are shown in figure 4.18 (A). In comparison to the temperature distribution at $t^* = 0$ where p_0^* is reached, as shown in figure 4.14 (A), the liquid temperature here, particularly in the vicinity of the free surface at T11 is higher than in the initial state for $t^* = 0$. Furthermore, it seems obvious that the thermal boundary layer is enlarged since the liquid temperature at T12 and T13 increase as well. Thus, the temperature

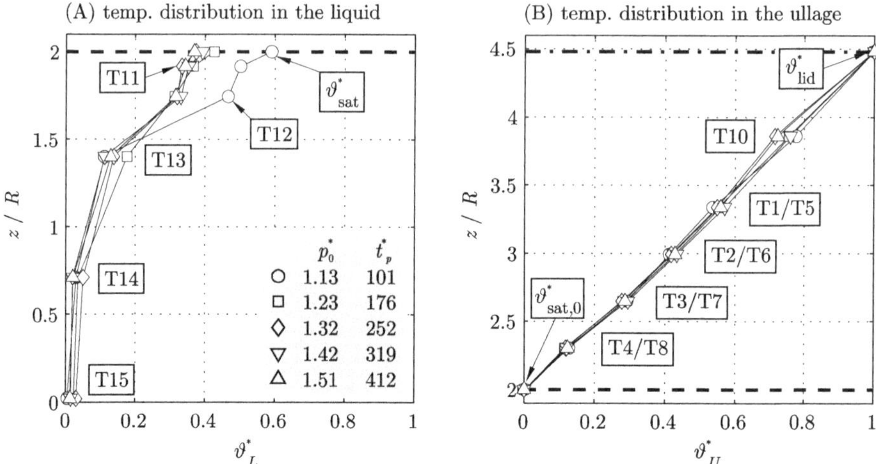

Figure 4.18: Temperature distribution for a GN_2 pressurized tank reaching the minimum pressure $p^* = p^*_{\mathrm{min}}$ in the liquid (A) and in the ullage (B). The dashed lines indicate the position of the free liquid surface, while the dash-dot line indicate the position of the tank lid. Open symbols correspond to the temperature data indicating different initial pressure levels. The solid lines merely emphasize the distribution to enhance the readability.

rise indicates the mixing effects in the thermal boundary layer, which is supported by the latent heat that is released by condensation. This heat dissipates through the liquid resulting in a more homogeneous temperature stratification in this region. This is confirmed by previous studies by LACAPERE *et al.* [39] and DAS & HOPFINGER [26]. The temperature at the free liquid surface is coupled to the tank pressure by the CLAUSIUS CLAPEYRON law. This relation is introduced in equation (2.82). While the surface temperature decreases during sloshing, the tank pressure decreases as well. The temperature distribution in the ullage after reaching the pressure minima is provided in figure 4.18 (B). Except for the lower strata at T4/T8 where the ullage temperature is slightly lower than observed for $t^* = 0$ in figure 4.14 (B), the temperature distribution in the ullage here remains stable without significant variations under the impact of sloshing.

On the next pages, the thermodynamic balance of the GN_2 pressurized system is provided. The dashed lines in figure 4.19 represent the control volume boundaries to balance the internal (liquid and ullage) and external (mass transport due to pressurization) energies. The total balance of energy during the GN_2 pressurization experiments is composed of two parts, the energy balance for the liquid and the energy balance for the ullage as shown in figure 4.19 yielding

$$dQ_{\mathrm{tot}} = dQ_L + dQ_U \,. \tag{4.38}$$

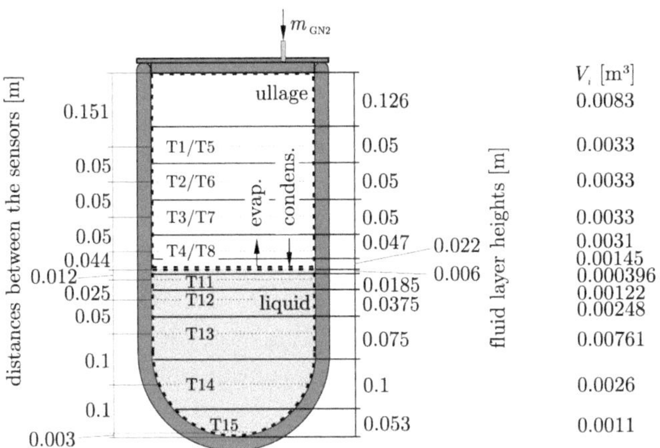

Figure 4.19: Control volumes inside of the tank for GN_2 pressurization in the liquid and in the ullage. The measures on the left side correspond to the distances between the temperature sensors, while the measures on the right side correspond to the fluid layer heights. The column on the far right side includes the according volume of each layer.

As already introduced for the self-pressurized system, the liquid can be considered as open system. Based on equation (2.14), the energy balance for the liquid is defined in general

$$dQ_L = dU_{e,L} + \sum m_{L,\text{out}}\, h_{e,L,\text{out}} - \sum m_{L,\text{in}}\, h_{e,L,\text{in}} \qquad (4.39)$$

where liquid can either enter (condensation) or leave (evaporation) the control volume via the free surface. As introduced in equation (2.17), the inner energy of the liquid can be expressed by the enthalpy, so that

$$dU_{e,L} = dH_{e,L} - p\, dV_L - V_L\, dp \qquad (4.40)$$

where $p\, dV_L = 0$ for constant volumes. Thus, the expression for the inner energy in the liquid simplifies

$$dU_{e,L} = dH_{e,L} - V_L\, dp\,. \qquad (4.41)$$

The change of the enthalpy is defined as difference between the enthalpy in state (I), e.g. start of pressurization, and state (II), e.g. end of pressurization, so that

$$dH_{e,L} = H_{e,L}^{(II)} - H_{e,L}^{(I)} = m_L^{(II)} h_{e,L}^{(II)} - m_L^{(I)} h_{e,L}^{(I)} \qquad (4.42)$$

with

$$m_L = \sum_k \varrho_{L,k}\,(\vartheta_{L,k})\, V_{L,k} \qquad (4.43)$$

defined as the mass of the liquid. As shown in figure 4.19, the liquid is subdivided into k layers based on the distances between the temperature sensors. Referring to the fluid database [41], the specific enthalpy $h_{e,L}(\vartheta_L)$ for each layer is determined from the measured temperature at sensor position T11 to T15. The liquid mass in each layer is determined by the liquid volume based on the fluid layer heights provided in figure 4.19 and the temperature depended liquid density $\varrho_L(\vartheta_L)$ taken from [41] as well. Thus, the enthalpy for the entire liquid volume reads

$$dH_{e,L} = \sum_k m_{L,k}^{(II)} h_{e,L,k}^{(II)} - \sum_k m_{L,k}^{(I)} h_{e,L,k}^{(I)} . \tag{4.44}$$

In the following, the energy balance for the liquid is defined in detail for the different phases; pressurization and sloshing. During pressurization with GN_2, the injected pressurant gas ($\vartheta_{GN2} = 293$ K) re-condenses inside the tank for some extent. This must be considered in the energy balance for the liquid as well, so that

$$dQ_{L,\text{press}} = dH_{e,L} - V_L \, dp - m_{\text{condens}} \left[h_{e,L,\text{condens}} + \Delta h_v \right] \tag{4.45}$$

where $m_{\text{condens}} \left[h_{e,L,\text{condens}} + \Delta h_v \right]$ is the converted vapor energy transferred into the liquid by vapor re-condensation during GN_2 pressurization. In this connection, $h_{e,L,\text{condens}}$ is the liquid enthalpy at surface level corresponding to the re-condensed vapor entering the liquid phase and Δh_v is the latent heat of evaporation released from the phase change from vapor to liquid. Equation (4.39) assumes that the mass m_{condens} enters the control volume as liquid. The conversion of the vapor mass into that liquid mass releases the latent heat of evaporation. This is considered in equation (4.45). Please note that is approach is equivalent to

$$dQ_{L,\text{press}} = dH_{e,L} - V_L \, dp - m_{\text{condens}} \, h_{e,U,\text{condens}} . \tag{4.46}$$

The enthalpy of the vapor mass entering the lower control volume is than subtracted from the upper control volume as shown later in equation (4.54). Furthermore, the determination of the mass of the re-condensed vapor during GN_2 pressurization is described afterwards in the energy balance for the ullage.

By neglecting evaporation due to the short time scale during sloshing, the energy balance in the liquid during the sloshing phase is similar to the pressurization phase, so that

$$dQ_{L,\text{slosh}} = dH_{e,L} - V_L \, dp - m_{\text{condens}} \left[h_{e,L,\text{condens}} + \Delta h_v \right], \tag{4.47}$$

again, where $m_{\text{condens}} \left[h_{e,L,\text{condens}} + \Delta h_v \right]$ is the energy transferred into the liquid by vapor condensation during sloshing. Analogous to the pressurization phase, $h_{e,L,\text{condens}}$ corresponds to the liquid enthalpy at surface level of the condensed mass to be added to the liquid. During the phase change from vapor to liquid, the latent heat Δh_v is released and therefore absorbed by the liquid. Similar to the pressurization phase, this approach is equivalent to

$$dQ_{L,\text{slosh}} = dH_{e,L} - V_L\, dp - m_{\text{condens}}\, h_{e,U,\text{condens}} \qquad (4.48)$$

where the enthalpy of the vapor mass entering the lower control volume is removed from the upper control volume and the released latent heat of evaporation is absorbed by the liquid.

Also the ullage is considered as open system where vapor can either enter (evaporation) or leave (condensation) the control volume via the free surface, while the pressurant gas (GN$_2$) entering the tank must be considered as well. Thus, the energy balance of the ullage is defined in general by

$$dQ_U = dU_{e,U} + \sum m_{\text{out}}\, h_{e,\text{out}} - \sum m_{\text{in}}\, h_{e,\text{in}} \qquad (4.49)$$

where the GN$_2$ injection during pressurization through the gas inlet is regarded in the source term $m_{\text{in}}\, h_{e,\text{in}}$ and (re)-condensation during external pressurization and sloshing is regarded in the sink term $m_{\text{out}}\, h_{e,\text{out}}$. As introduced in equation (2.17), the inner energy of the ullage can be expressed by the enthalpy yielding

$$dU_{e,U} = dH_{e,U} - p\, dV_U - V_U\, dp\,, \qquad (4.50)$$

again, where $p\, dV_U = 0$ for constant volumes. As well in the ullage, the change of the enthalpy is defined as difference between the enthalpy in state (I) and state (II), so that

$$dH_{e,U} = H_{e,U}^{(II)} - H_{e,U}^{(I)} = m_U^{(II)}\, h_{e,U}^{(II)} - m_U^{(I)}\, h_{e,U}^{(I)}\,. \qquad (4.51)$$

The mass of the ullage gas (vapor) m_U is determined by

$$m_U = \sum_k \varrho_{U,k}\,(\vartheta_{U,k})\, V_{U,k}\,. \qquad (4.52)$$

The maximum error of the approximation given in equation (4.52) is about 5% with respect to a linear temperature stratification of the ullage with ϑ_{sat} at the free surface and ϑ_{lid} at the tank lid as shown in figure 4.14 (B).

As shown in figure 4.19, also the ullage is subdivided into k layers based on the distances between the temperature sensors. Again, the specific enthalpy $h_{e,U}(\vartheta_U)$ for each layer is determined from the measured temperature at sensor position T1/T5 to T4/T8 taken from [41]. The gas mass in each layer is determined by the gas volume provided in figure 4.19 and the temperature depended gas density $\varrho_U(\vartheta_U)$ taken from [41] as well. Thus, the enthalpy for the entire ullage volume reads

$$dH_{e,U} = \sum_k m_{U,k}^{(II)}\, h_{e,U,k}^{(II)} - \sum_k m_{U,k}^{(I)}\, h_{e,U,k}^{(I)}\,. \qquad (4.53)$$

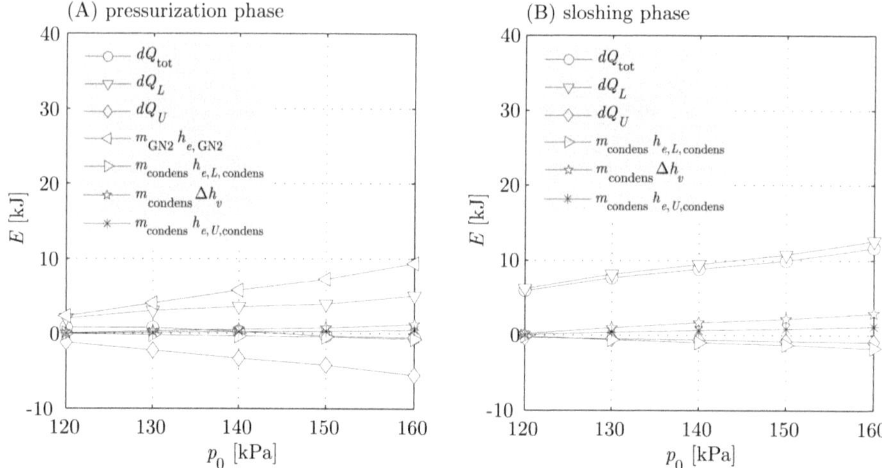

Figure 4.20: Energy balance of GN_2 pressurized systems for the pressurization phase (A) and for the sloshing phase (B). The solid lines connect the data points to increase the readability.

In the following, the energy balance for the ullage is defined in detail for the different phases; pressurization and sloshing. For pressurization, the energy balance in the ullage reads

$$dQ_{U,\,\mathrm{press}} = dH_{e,\,U} - V_U\,dp - m_{\mathrm{GN2}}\,h_{e,\,\mathrm{GN2}} + m_{\mathrm{condens}}\,h_{e,\,U,\,\mathrm{condens}} \tag{4.54}$$

where the specific enthalpy of the pressurant gas is $h_{e,\,\mathrm{GN2}} = 303.9\,\mathrm{kJ\,kg^{-1}}$ for a gas temperature of $\vartheta_{\mathrm{GN2}} = 293\,\mathrm{K}$. The gaseous nitrogen mass m_{GN2} to pressurize the tank is provided in table A.3 in the appendix. Furthermore, $m_{\mathrm{condens}}\,h_{e,\,U,\,\mathrm{condens}}$ is the energy leaving the ullage due to re-condensation during GN_2 pressurization, while $h_{e,\,U,\,\mathrm{condens}}$ corresponds to the ullage enthalpy at surface level. The mass of the re-condensed vapor during GN_2 pressurization of the tank is determined by integrating the respective vapor mass over all gas layer (see figure 4.19), so that

$$m_{\mathrm{condens}} = m_U^{(II)} - m_U^{(I)} = \sum_k m_{U,\,k}^{(II)} - \sum_k m_{U,\,k}^{(I)}, \tag{4.55}$$

while state (I) corresponds to the start of pressurization and state (II) corresponds to the end of pressurization. The vapor mass in each layer is determined by

$$m_{U,\,k} = \varrho_{U,\,k}\left(\vartheta_{U,\,k}\right) V_{U,\,k}. \tag{4.56}$$

During the sloshing phase, all inlets and outlets are closed. Mass can only leave the ullage control volume via the free surface in terms of phase change (condensation). Thus, the energy balance for the ullage during the sloshing phase reads

$$dQ_{U,\text{slosh}} = dH_{e,U} - V_U\,dp + m_{\text{condens}}\,h_{e,U,\text{condens}}\,. \qquad (4.57)$$

Again here, $m_{\text{condens}}\,h_{e,U,\text{condens}}$ is the energy released by vapor condensation during sloshing. Analogous to the pressurization phase, $h_{e,U,\text{condens}}$ corresponds to the ullage enthalpy at surface level, while the condensed mass is determined according to equation (4.55).

The energy balance for the pressurization phase is shown in figure 4.20 (A). The energy induced by the pressurant gas $m_{\text{GN2}}\,h_{e,\text{GN2}}$ has the highest impact on the total energy as indicated by the $\triangleleft$ symbols. Contrary to the previous case, the change of the energy in the liquid corresponding to dQ_L is about 2 times smaller than during self-pressurization. This can be explained by the strong thermal gradient in the liquid observed after GN_2 pressurization, while the thermal boundary layer during self-pressurization is of deeper penetration allowing more liquid to heat up. The external pressurization leads to a negative balance in the ullage (heat is emitted) since the warm pressurant gas is added to the ullage, while some vapor re-condenses in the vicinity of the free surface. The re-condensation is indicated by $\triangleright$ symbols on the liquid side and by $*$ symbols on the ullage side. The difference between these values divided by the re-condensed mass fraction m_{condens} gives the latent heat of evaporation Δh_v released by the phase change. Values for the pressurization phase and the sloshing phase are presented in table A.29 and table A.30 in the appendix.

The energy balance for the GN_2 pressurized system during the sloshing phase is provided in figure 4.20 (B) as function of the initial pressure. As already observed for self-pressurization, the process is driven by condensation initiated by mixing effects in the thermal boundary layer under

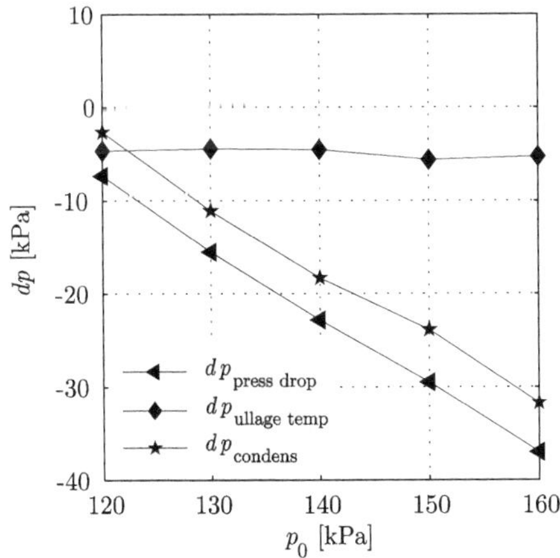

Figure 4.21: Composition of the pressure drop for the GN_2 pressurized systems showing the pressure change in the ullage. The solid lines connect the data points to increase the readability.

the free liquid surface where the surface temperature and therefore the saturation temperature are decreasing during sloshing. Due to the mixing and absorbing the latent heat, the averaged temperature of the liquid slightly increases leading to an increase of the energy in the liquid as indicated by the $\triangledown$ symbols. The increase is similar (slightly smaller) than the dQ_L determined during the sloshing phase after self-pressurization. As well as after self-pressurization, the energy difference in the ullage after GN_2 pressurization is negative implying condensation and the ullage cooling-down caused by the sloshing liquid as indicated by the $\Diamond$ symbols. The energy that is added to the liquid due to condensation is indicated by $\triangleright$ symbols while the energy leaving the ullage is indicated by $*$ symbols.

Analogous to the self-pressurization experiments, the pressure drop that occurs under the impact of sloshing is superimposed by two different effects; the ullage cooling-down as well as condensation effects caused by the decrease of the surface temperature. This is shown in figure 4.21 where the full $\blacktriangleleft$ symbols correspond to the total pressure drop $dp_{\text{press drop}}$ measured by means of the pressure sensor inside the tank. By knowing the mean gas temperature and the mean mass distribution from the ullage gas densities, the pressure loss fraction caused by the temperature decrease of the ullage gas is determined by means of equation (4.35) and equation (4.36). The data for $dp_{\text{ullage temp}}$ is indicated by the full $\blacklozenge$ symbols. For all considered initial pressures, the pressure drop due to the ullage temperature decrease is approximately constant and therefore not depended on p_0. The pressure drop due to condensation is thus

$$dp_{\text{condens}} = dp_{\text{press drop}} - dp_{\text{gas temp}} \tag{4.58}$$

indicated by the full $\bigstar$ symbols. Except for the smallest initial pressure, the fraction of condensation is about 90% of the total pressure loss, while the fraction of ullage temperature decrease is about 10% for the GN_2 pressurized system.

4.3.4 Helium Pressurization

Helium (GHe) pressurized propellant tanks must be considered as two-species systems where the pressurant gas corresponds to a non-condensable inert gas in a condensable propellant vapor environment. Other than in the previous test cases where one-species systems are considered, this test case represents a convenient method to pressurize propellant tanks including the advantage of reducing phase change effects such as condensation at the liquid/vapor interface that typically occur in one-species systems under the impact of sloshing. Actual examples for the application include Ariane 5 ESC-A, the Space Shuttle and the Centaur upper stage, to name only a few. An illustration of the helium pressurized test system is provided in figure 4.22 showing the dewar tank including its instrumentation. The helium pressurization is realized by feeding the pressurant gas from an external GHe bottle that is connected to the test tank. The

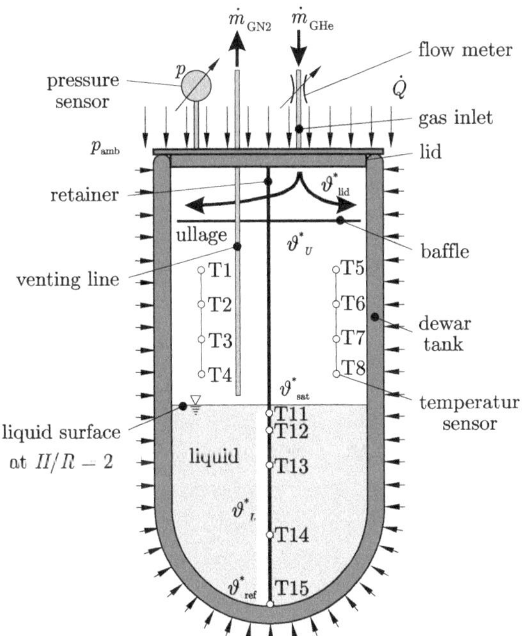

Figure 4.22: Principle for pressurizing the dewar tank with helium (GHe) including the instrumentation in the liquid and in the ullage. The heat flux from outside the tank is indicated by small arrows on the light gray area. The locations of the temperature sensors are marked by the open dots. Qualitative temperature distributions in the liquid and in the ullage are indicated by light gray areas inside the tank. Helium is fed into the tank through the gas inlet at the tank lid, while GN_2 from the surface region can be flushed out through the venting line. Therefore, the venting line is positioned close above the liquid surface (≈ 5 mm). The helium flow rate of $\dot{m}_{GHe} = 0.02$ g s^{-1} is controlled by a flow meter.

Table 4.5: Molar helium fraction $\chi_{\mathrm{GHe},f}$ and $\chi_{\mathrm{GHe},f+p}$ as well as the helium mass fraction $x_{\mathrm{GHe},f}$ and $x_{\mathrm{GHe},f+p}$ after the flushing phase t_f^* and after the entire pressurization phase t_{f+p}^*. The alphanumeric labeling corresponds to the positions in figure 4.23. The characteristic time is defined by $\tau = 1/f = 0.714$ s.

	a	b	c	d	e	f	g	h
$t_f^* = t_f/\tau$	25.2	67.2	109.2	151.3	193.2	235.3	277.3	319.3
$\chi_{\mathrm{GHe},f}$ [mol/mol]	0.06	0.17	0.28	0.38	0.49	0.59	0.69	0.79
$x_{\mathrm{GHe},f}$ [kg/kg]	0.01	0.03	0.05	0.08	0.12	0.17	0.24	0.35
	a'	b'	c'	d'	e'	f'	g'	h'
$t_{f+p}^* = t_{f+p}/\tau$	203.1	240.9	277.3	316.5	358.5	317.9	418.8	458.0
$\chi_{\mathrm{GHe},f+p}$ [mol/mol]	0.38	0.46	0.52	0.60	0.68	0.73	0.79	0.87
$x_{\mathrm{GHe},f+p}$ [kg/kg]	0.08	0.11	0.14	0.17	0.23	0.28	0.23	0.48

pressurant mass flow rate for all helium experiments performed for this work corresponds to $\dot{m}_{\mathrm{GHe}} = 0.02$ g s^{-1}, while the pressurant temperature is $\vartheta_{\mathrm{GHe}} \approx 293$ K. The helium pressurant is brought into the system through the gas inlet integrated in the tank lid. The gas inlet is located 100 mm off the tank center. Furthermore, the tank is equipped with a baffle to prevent that the pressurant gas jet directly hits the liquid surface. In a first approximation, it is assumed that the helium is approximately evenly distributed along the perimeter of the baffle. In consequence, this conserves the thermal stratification in the ullage during pressurization, which is important to ensure repeatable initial conditions in the tank. To test the influence of the helium concentration on the tank pressure development, experiments are performed for a constant initial pressure of $p_0^* = 1.32$, while varying the helium flushing duration of the ullage. As already determined in the previous test cases, the heat entering through the tank walls is measured to be approximately $\dot{Q}_{\mathrm{heat}} = 5.9$ W.

Experiments are started by filling the pre-cooled tank with LN$_2$ up to the required fill level of $H/R = 2$. After closing the tank lid, the system can be considered as one-species system where the ullage represents a 100% GN$_2$ atmosphere. In order to reduce the existing nitrogen vapor first, the tank is purged with helium before the pressurization is started. Therefore, helium from an external bottle is brought into the tank through the gas inlet at the tank lid, while displaced GN$_2$ can escape near the liquid surface through the venting line into the environment. During this procedure, the tank pressure increases only slightly. The venting line inlet is located approximately 5 mm above the liquid surface to ensure that only GN$_2$ can be extracted from

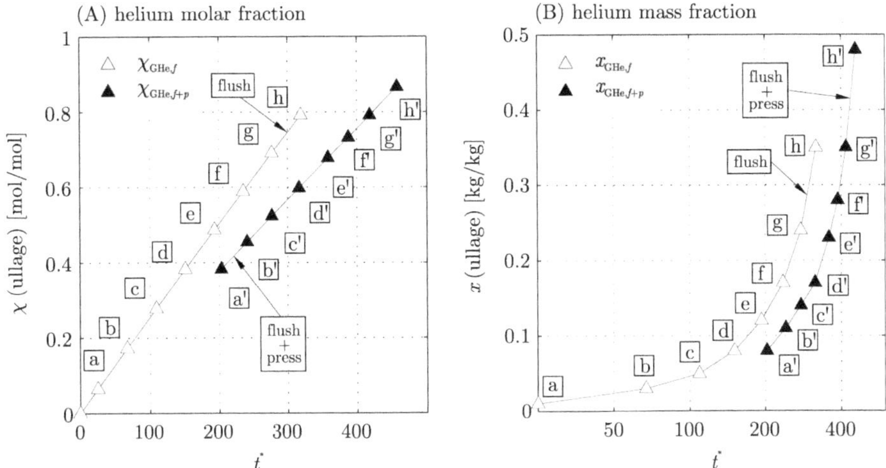

Figure 4.23: (A) The helium molar fraction χ_{GHe} in the ullage after the flushing ($\triangle$) and pressurization phase ($\blacktriangle$). (B) The helium mass fraction x_{GHe} in the ullage after the flushing ($\triangle$) and pressurization phase ($\blacktriangle$). The data is based on table 4.5. The alphanumerical labeling corresponds to the different flushing times.

the ullage[3]. The tank is flushed for various time periods to test the influence of the helium concentration in the ullage on the impact of the pressure drop.

By reaching t_f^*, the GN$_2$ venting line is closed and the tank is pressurized with GHe up to the required initial pressure of $p_0^* = 1.32$. Expressed by different letters, the flushing times t_f^* and the pressurization times t_{f+p}^* as well as the corresponding helium mole fraction χ_{GHe} and the helium mass fraction x_{GHe} are listed in table 4.5. The concentrations are determined from the data provided in table A.4 in the appendix. Assuming that only GN$_2$ is vented during the flushing period, the according helium molar fraction χ_{GHe} is determined from the known amount of substances and volumes, so that

$$\chi_{GHe} = \frac{n_{GHe}}{n_U} \qquad \chi_{GN2} = \frac{n_{GN2}}{n_U} \tag{4.59}$$

with the amount of helium and nitrogen gas

$$n_{GHe} = \frac{m_{GHe}}{M_{GHe}} \qquad n_{GN2} = \frac{m_{GN2}}{M_{GN2}} \tag{4.60}$$

and the total amount of substance in the ullage $n_U = n_{GHe} + n_{GN2}$. Based on a mean ullage temperature of $\bar{\vartheta}_U \approx 185$ K and a tank pressure of 100 kPa, the mean density of GHe is $\bar{\varrho}_{GHe} = 0.26$ kg m^{-3}, while the mean density of GN$_2$ is $\bar{\varrho}_{GN2} = 1.82$ kg m^{-3}. The mass of helium

---

[3]Of course, for very long flushing duration $t_f^* > 350$ helium might be flushed as well.

brought into the tank during flushing is $m_{\mathrm{GHe}} = \dot{m}_{\mathrm{GHe}}\,t_f$ provided in table A.4 in the appendix. Thus, the ullage volume assigned by helium is $V_{\mathrm{GHe}} = m_{\mathrm{GHe}}/\bar{\varrho}_{\mathrm{GHe}}$, while the ullage volume assigned by gaseous nitrogen is $V_{\mathrm{GN2}} = V_U - V_{\mathrm{GHe}}$. Furthermore, the gaseous nitrogen mass after flushing is $m_{\mathrm{GN2}} = V_{\mathrm{GN2}}\,\bar{\varrho}_{\mathrm{GN2}}$. For pressurization, the mean density of gaseous helium is $\bar{\varrho}_{\mathrm{GHe}} = 0.29\ \mathrm{kg\,m^{-3}}$, while the mean density of the gaseous nitrogen is $\bar{\varrho}_{\mathrm{GN2}} = 2.03\ \mathrm{kg\,m^{-3}}$. The gaseous helium molar mass is $M_{\mathrm{GHe}} = 0.004\ \mathrm{kg\,mol^{-1}}$ and the gaseous nitrogen molar mass is $M_{\mathrm{GN2}} = 0.028\ \mathrm{kg\,mol^{-1}}$. The mass fractions for substances in the ullage give

$$x_{\mathrm{GHe}} = \frac{\chi_{\mathrm{GHe}}\,M_{\mathrm{GHe}}}{\bar{M}_U} \qquad x_{\mathrm{GN2}} = \frac{\chi_{\mathrm{GN2}}\,M_{\mathrm{GN2}}}{\bar{M}_U} \tag{4.61}$$

where

$$\bar{M}_U = \chi_{\mathrm{GHe}}\,M_{\mathrm{GHe}} + \chi_{\mathrm{GN2}}\,M_{\mathrm{GN2}} \tag{4.62}$$

is the average molar mass of the ullage. All quantities are provided in table A.5 and A.6 in the appendix. It is observed that for increasing flushing time the pressurization time decreases due to the enriched helium atmosphere in the ullage. For the same amount of mass (helium and nitrogen), the specific volume of GHe (reciprocal of the density) is higher than the one of $\mathrm{GN_2}$. In case of constant volume, the required initial pressure is reached in a shorter period for increasing flushing times. The helium molar fractions χ_{GHe} for the flushing and the pressurization phase versus time are plotted in figure 4.23 (A), while the helium mass fraction is shown in figure 4.23 (B). Open triangle symbols correspond to the helium concentration in the ullage after flushing $\chi_{\mathrm{GHe},f}$, while the full triangle symbols correspond to the helium concentration after flushing and pressurization $\chi_{\mathrm{GHe},f+p}$.

As a matter of fact, the concentration of gaseous nitrogen strongly increases in the vicinity of the free liquid surface leading to a strong concentration gradient in this region. Here, the liquid surface that is at saturation temperature represents a $\mathrm{GN_2}$ source that further feeds the ullage with nitrogen vapor. After pressurization, the concentration in the vicinity of the liquid surface is determined by measuring the liquid temperature directly at the liquid surface. This is realized by adding a temperature sensor that is fixed on a small float near the center of the free liquid surface during flushing and pressurization. The sensor slightly touches the liquid in order to measure only the surface temperature. Note that this sensor is ineffective, while the surface is in motion[4]. Satisfying the CLAUSIUS CLAPEYRON equation (2.82), the nitrogen partial pressure p_{GN2} is determined from the liquid surface temperature $\vartheta_{\mathrm{sat},0}$ provided in table A.7 in the appendix. Thus, the nitrogen concentration of the ullage in the vicinity of the free surface is determined by

$$\chi_{\mathrm{GN2}} = \frac{p_{\mathrm{GN2}}(\vartheta_{\mathrm{sat},0})}{p}, \tag{4.63}$$

---

[4]The float carrying the sensor is removed for sloshing experiments.

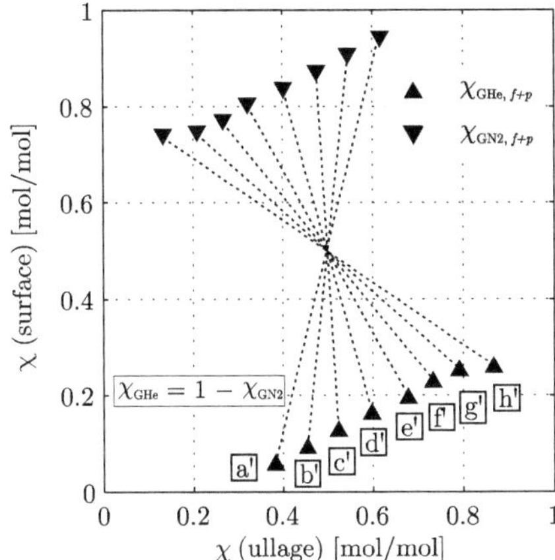

Figure 4.24: The helium concentration data in the ullage after flushing and pressurization (abscissa) is correlated to measurement data of the helium concentration directly at the liquid surface (ordinate). The helium concentration is derived from the nitrogen partial pressure corresponding to the surface temperature ϑ_{sat}. The dashed lines indicate the coupled GHe and GN_2 concentrations according to equation (4.64).

while the helium partial pressure can be determined by

$$\chi_{\mathrm{GHe}} = 1 - \chi_{\mathrm{GN2}} \,. \tag{4.64}$$

The results are shown in figure 4.24 where the species concentrations χ_{GN2} and χ_{GHe} at the free liquid surface are plotted versus the average concentrations in the ullage determined by the known mass fractions during flushing and pressurization. While the helium concentration increases for increasing flushing times, of course the nitrogen concentration must decrease. The nitrogen concentration at the surface decreases by only approximately 20%, while the nitrogen concentration in the entire ullage decreases by approximately 60% for increasing helium flushing times. This results allow the conclusion that it exist a strong concentration gradient in the vicinity of the free liquid surface where the helium concentration is about a factor 4 smaller than the average concentration in the ullage. According to BAEHR & STEPHAN [12], the appearance of an inert gas implies the existence of a partial pressure difference at the phase interface that drives the vapor to diffuse through the non-condensable pressurant. This partial pressure difference strongly depends on the helium concentration. For the cryogenic system considered here, the GN_2 partial pressure increases in the vicinity of the phase interface according to the saturation temperature of the liquid. Hence, the partial pressure of the helium pressurant must decrease to satisfy the total tank pressure according to DALTONs law yielding

$$p = p_{\mathrm{GN2}} + p_{\mathrm{GHe}} \, . \tag{4.65}$$

Since all helium pressurization experiments are conducted with an initial pressure of $p_0^* = 1.32$, the dissolution of helium in LN_2 is estimated as well. The solubility of a gaseous species (GHe) in a different liquid species (LN_2) is for a dilute concentration defined by HENRYs law. This relation yields for the certain case as considered here

$$p_{\mathrm{GHe}} = k_H \left(\vartheta_L \right) \psi_{\mathrm{GHe}} \tag{4.66}$$

where p_{GHe} is the helium partial pressure, ψ_{GHe} is the concentration of the dissolved helium in the liquid and k_H is the HENRY constant that depends on the solute, the solvent and the liquid temperature ϑ_L. For the highest helium concentration, $k_H \approx 423800$ kPa. The concentration of helium dissolved in LN_2 can be determined by an empiric correlation given by VAN DRESAR [28] that yields

$$\psi_{\mathrm{GHe}} = 1.383\mathrm{E}{-}10 \left(p_0 \, [\mathrm{MPa}] - p_{\mathrm{sat}} \, [\mathrm{MPa}] \right)^{0.99} \vartheta_L^{3.82} . \tag{4.67}$$

For the highest flushing duration, $p_0 = 0.14$ MPa is the tank pressure after helium pressurization and $p_{\mathrm{sat}} = 0.104$ MPa is the saturation pressure corresponding to the liquid temperature $\vartheta_L = 77.6$ K. Note that the pressure quantities in equation (4.67) must be provided in [MPa]. The solubility of gas dissolved in the liquid is basically a function of the liquid temperature and therefore of the tank pressure as well. For the highest helium concentration in the ullage, the concentration of dissolved helium in LN_2 gives $\psi_{\mathrm{GHe}} = 8.53\mathrm{E}{-}5$ mol mol^{-1}. Thus, the helium concentration ratio gives $\psi_{\mathrm{GHe}}/\chi_{\mathrm{GHe}} \ll 1$ and can therefore be neglected. Sloshing might increase the mass transport of dissolved helium for some extent, but regarding the exposure time in the order of minutes, this can be neglected as well.

After the required initial pressure of $p_0^* = 1.32$ is reached, the periodic excitation is initiated by starting the engine. The liquid starts to slosh in the first asymmetric sloshing (lateral) mode. As described in chapter 4.3.2, the lateral oscillation is defined by a frequency ratio of $\eta_{11} = 0.78$ and an amplitude ratio of $y_A/R = 0.069$. During the experimental run, which includes the flushing phase, the pressurization phase and the sloshing phase, the tank pressure as well as the temperatures in the liquid and in the ullage are logged.

The temperature distribution after helium flushing and helium pressurization for the required initial pressure p_0^* is shown in figure 4.25. Considering the dimensionless formulation introduced in equation (4.9), the temperature distribution in the liquid is plotted in figure 4.25 (A) for the scaled tank height z/R. Likewise as it appears during the GN_2 pressurization, the highest temperature gradient in the liquid is observed in the vicinity of the free surface. Assuming a mixture of ideal gases in the ullage, the free surface must have saturation temperature ϑ_{sat} corresponding to the nitrogen partial pressure p_{GN2} of the gas mixture in the vicinity of the liquid surface [11]. While the liquid temperature is uniformly $\vartheta_L^* \approx 0$ distributed in the entire

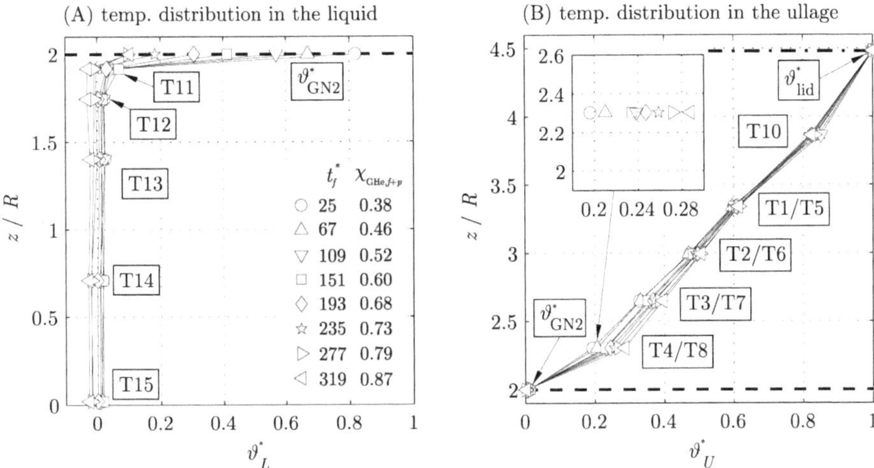

Figure 4.25: Temperature distribution after GHe pressurization reaching the required initial pressure p_0^*. The dashed lines indicate the position of the free liquid surface, while the dash-dot line indicate the position of the tank lid. Open symbols correspond to the temperature data indicating different helium concentration levels after pressurization. The solid lines merely emphasize the distribution to enhance the readability. The inset in (B) highlights the pressure depended temperature distribution at sensor T4/T8.

liquid for $0 \leq z/R \leq 2$ after filling the tank to the required fill level, obviously the liquid temperature in the bulk does not change significantly during helium pressurization. This can be explained by the fact that during pressurization only the helium partial pressure increases.

The temperature distribution in the ullage along the scaled tank height z/R is provided in figure 4.25 (B). As already shown for the previous test cases considering self-pressurization and GN_2 pressurization, the ullage temperature between T4/T8 and T1/T5 is linearly stratified. The ullage temperature between T4/T8 and T2/T6 shows a clear dependency on the flushing time and therefore on the helium concentration in the ullage as indicated in the inset of figure 4.25 (B). It is clearly shown that the ullage temperature in the lower strata is coupled to the helium concentration in the ullage. The temperature of the pressurant gas is about $\vartheta_{GHe} = 293$ K, but this is supposed to be insufficient to explain the temperature variation in the ullage. Experiments where colder pressurant gas is utilized show similar characteristics.

The pressure development of the helium pressurized systems is provided in figure 4.26. The open symbols correspond to different flushing times t_f and therefore to different helium concentrations in the ullage χ_{GHe}. The pressurization phase including the flushing phase is shown in figure 4.26 (A) for $t^* < 0$. In here, the different gradients allow the assumption that the

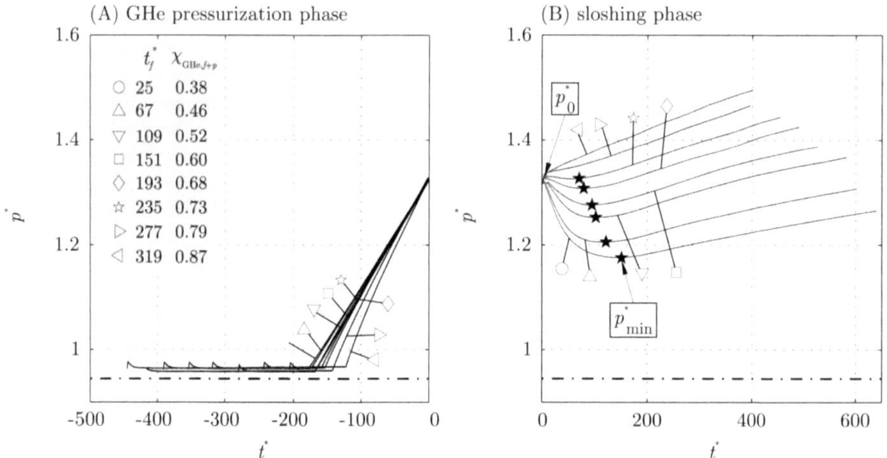

Figure 4.26: Pressure development for experiments with helium pressurization. The pressurization phase corresponding to $t^* \leq 0$ is provided in (A), while the sloshing phase corresponding to $t^* \geq 0$ is provided in (B). The open symbols indicate different helium concentration levels corresponding to different flushing times, while the solid lines correspond to the experimental pressure data. The pressure minima for the according helium flushing times t_f^* are indicated by $\bigstar$ symbols. The dash-dot lines correspond to the norm pressure level.

duration to pressurize the tank up to the required initial pressure p_0^* strongly depends on the helium concentration in the ullage. Thus, the steepest ramp corresponds to the highest helium concentration. For increasing helium concentration, the LN_2 saturation temperature decreases provoking further evaporation and thus a faster increase of the tank pressure.

The required initial pressure of $p_0^* = 1.32$ is reached at $t^* = 0$. Then the tank is excited for $t^* > 0$. The pressure development during the sloshing phase shown in figure 4.26 (B) confirms the expectation that the pressure drop strongly depends on the helium concentration in the tank. The magnitude of the pressure drop $p_0^* - p_{min}^*$ decreases for increasing helium concentrations. The pressure minima p_{min}^* are indicated by $\bigstar$ symbols. It is remarkable that the characteristic pressure drop even disappears for helium concentrations $\chi_{GHe,f+p} > 0.73$.

After passing the pressure minimum at p_{min}^*, the tank pressure re-increases. Other than observed in the previous test cases considering one-species systems, the pressure gradients here are higher and obviously depend on the helium concentrations whereas the highest helium concentration corresponds to the steepest pressure gradient after passing p_{min}^* as shown in figure 4.26. Again, this can be explained by the decreasing nitrogen partial pressure for increasing helium concentration, which provokes enhanced evaporation.

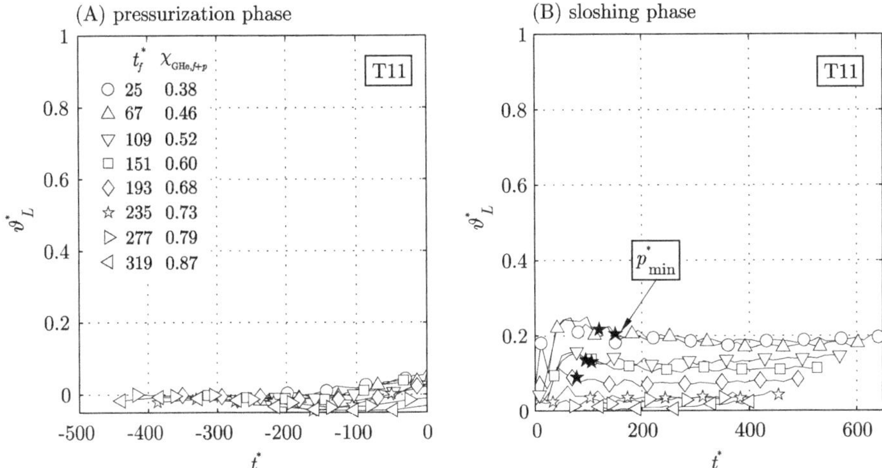

Figure 4.27: Temperature development in the liquid at sensor position T11 for GHe pressurization during the pressurization phase (A) and during the sloshing phase (B). The open symbols indicate different helium concentration levels due to different flushing times to enhance the readability, while the solid lines correspond to the experimental temperature data. The ★ symbols indicate the pressure minima p_{min}^*.

The temperature development below the free liquid surface is provided in figure 4.27 where the data corresponds to the liquid temperatures measured at sensor T11. The pressurization phase is shown in figure 4.27 (A) for $t^* < 0$. It is observed that the pressurization of the tank has only a small impact on the liquid temperature at T11. As well here, the impact of the helium concentration on the temperature development under the free liquid surface is less significant. With respect to the nitrogen partial pressure, the impact on the liquid temperature even decreases for increasing helium concentrations.

Reaching the initial pressure $p_0^* = 1.32$, the excitation is started at $t^* = 0$. The temperature development in the liquid at T11 during the sloshing phase is shown in figure 4.27 (B). With the initiation of liquid sloshing, the temperature at T11 rapidly increases as a result of the mixing effects and the dissipation of the released latent heat due to condensation. After a while, the pressure minima p_{min}^* are reached indicated by ★ symbols. Here, the thermal equilibrium in the liquid/vapor transition is re-established. The liquid temperature in the boundary layer in the vicinity of the free liquid surface and therefore the tank pressure re-increase.

The temperature development in the ullage is provided in figure 4.28 showing the averaged temperatures for the sensor couples T1/T5 to T4/T8. The ullage temperature development during the pressurization phase for $t^* < 0$ is shown in figure 4.28 (A). As well as observed at the beginning of the pressurization, the formation of the temperature stratification occurs immediately after refilling the tank up to the required fill level. When the data logging is started,

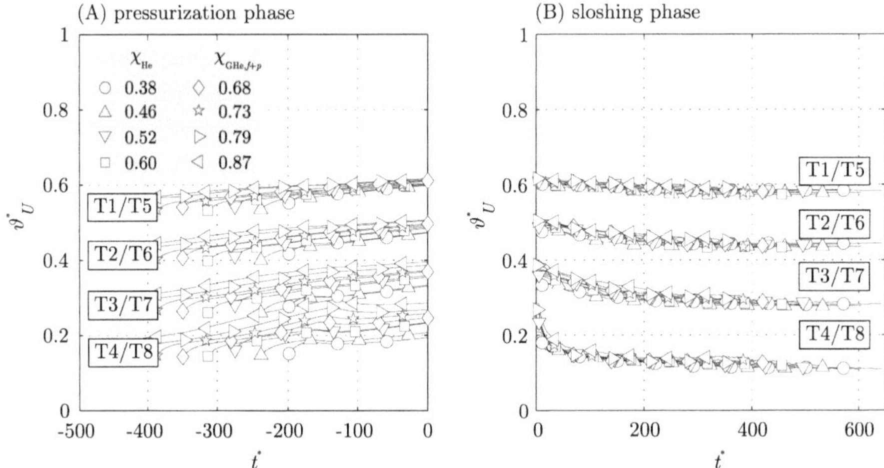

Figure 4.28: Temperature development in the ullage for helium pressurization experiments during the pressurization phase (A) and during the following sloshing phase (B). The ullage temperatures are averaged for the corresponding sensor pairings. The symbols indicate different helium concentration levels due to different flushing times to enhance the readability, while the solid lines correspond to the scaled measurement data.

the linear stratification in the ullage is already existent. Further temperature increase during helium pressurization can be traced back to the additional heat input due to pressurization with warmer gas. However, at the end of the pressurization phase at $t^* = 0$, the ullage is clearly stratified showing similar thermal stratifications that only varies in magnitude for all considered initial pressures. Particularly in the lower strata at T4/T8 and T3/T7 the ullage temperature development shows a clear dependency on the helium concentration, where the warmer ullage is observed for the higher helium concentration and thus for the longer flushing duration.

After starting the excitation, which is shown in figure 4.28 (B), only the layers in the lower regions at T4/T8 and T3/T7 are affected by the sloshing liquid. Here, the ullage temperature decreases driven by the sloshing liquid, so that the ullage cools down, while the tank pressure decreases. In the upper regions on the level of T2/T6 and T1/T5, the sloshing liquid does not have any significant influences on the temperature development in the ullage. Overall, the thermal stratification in the ullage is preserved, while only the temperature close to the sloshing liquid surface decreases.

The temperature distribution for the minimum pressure at $p^* = p^*_{min}$ is provided in figure 4.29. The temperature in the liquid is plotted versus scaled tank height z/R as shown in figure 4.29 (A). As well as observed in the previous test cases considering self-pressurization

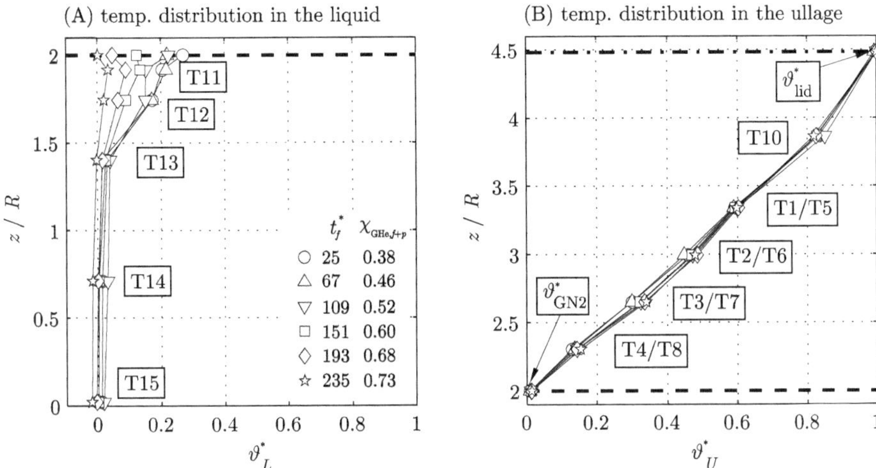

Figure 4.29: Temperature distribution for the GHe pressurized tank reaching the minimum pressure $p^* = p^*_{\min}$. The dashed lines indicate the position of the free liquid surface, while the dash-dot line indicate the position of the tank lid. Open symbols correspond to the temperature data indicating different helium concentration levels after pressurization. The solid lines merely emphasize the distribution to enhance the readability.

and GN_2 pressurization, the temperature stratification after reaching $p^* = p^*_{\min}$ appears more homogeneous. While the highest temperature gradient in the liquid is observed in the vicinity of the free surface after reaching the required initial pressure of $p^* = p^*_0$ as shown in figure 4.25 (A), this is assumed to be not the case during sloshing when the minimum pressure is reached. Although, the current setup does not allow the measurement of the surface temperature during sloshing, the surface temperature can be approximated by the liquid temperature at T11. This is justified by the result for GN_2 pressurization shown in figure 4.18 (A). Due to mixing effects in the thermal boundary, the liquid temperature at the surface, at T11 and at T12 appears approximately similar. Nevertheless, since only GN_2 is affected by condensation above the free surface, it is assumed that the helium partial pressure persists constant. Variations of the concentration gradient due to liquid sloshing would explain the spreading of the data at the liquid surface. However, the liquid temperature of the liquid bulk does not change significantly during sloshing.

The according temperature distribution in the ullage is provided in figure 4.29 (B). The temperature is still linearly stratified. In comparison to figure 4.25 (B) where the ullage temperatures at T4/T8 and T3/T7 show a strong dependency on the helium concentration, the ullage temperatures for $p^* = p^*_{\min}$ appear uniform and independent from the according helium concentration. Similar temperature distributions during sloshing are observed for self-pressurization and GN_2 pressurization. This might imply that the temperature stratification in the ullage particularly in lower regions is predominantly affected by the colder sloshing liquid surface and only for a smaller extent by the pressurant gas.

On the next pages, the thermodynamic balance of the GHe pressurized system is provided. The dashed lines in figure 4.30 represent the control volume boundaries to balance the internal (liquid and ullage) and external (mass transport due to flushing and GHe pressurization) energies inside the tank. The total balance of energy during the GHe pressurization experiments is composed of two parts, the energy balance for the liquid and the energy balance for the ullage yielding

$$dQ_{\text{tot}} = dQ_L + dQ_U \,. \tag{4.68}$$

As already introduced before, the liquid can be considered as open one-species system with phase change at the free surface assuming that dissolved helium can be neglected. Based on equation (2.14), the energy balance in the liquid can be rewritten in general by

$$dQ_L = dU_{e,L} + \sum m_{L,\text{out}}\, h_{e,L,\text{out}} - \sum m_{L,\text{in}}\, h_{e,L,\text{in}} \tag{4.69}$$

where mass can either enter (condensation) or leave (evaporation) the control volume via the free surface. According to equation (2.17), the inner energy of the liquid can be expressed also by the enthalpy, so that

$$dU_{e,L} = dH_{e,L} - p\,dV_L - V_L\,dp \tag{4.70}$$

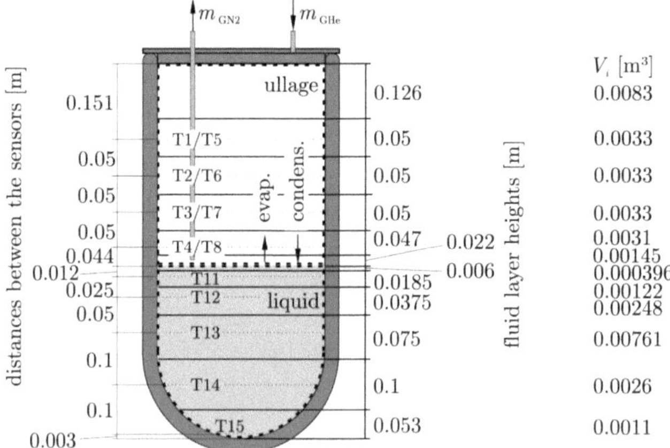

Figure 4.30: Control volumes inside of the tank for GHe pressurization in the liquid and in the ullage. The measures on the left side correspond to the distances between the temperature sensors, while the measures on the right side correspond to the fluid layer heights. The column on the far right side include the according volume of each layer.

where $p\,dV_L = 0$ for constant volumes. Thus, the expression for the inner energy in the liquid simplifies

$$dU_{e,L} = dH_{e,L} - V_L\,dp\,. \tag{4.71}$$

The change of the enthalpy is defined as the difference between the enthalpy in state (I), e.g. start of pressurization, and state (II), e.g. end of pressurization, so that

$$dH_{e,L} = H_{e,L}^{(II)} - H_{e,L}^{(I)} = m_L^{(II)}\,h_{e,L}^{(II)} - m_L^{(I)}\,h_{e,L}^{(I)} \tag{4.72}$$

with

$$m_L = \sum_k \varrho_{L,k}\,(\vartheta_{L,k})\,V_{L,k} \tag{4.73}$$

as the mass of the liquid. As shown in figure 4.30, the liquid is subdivided into k layers based on the distances between the temperature sensors. Referring to the fluid database [41], the specific enthalpy $h_{e,L}(\vartheta_L)$ for each layer is determined from the measured temperature at sensor position T11 to T15. The liquid mass in each layer is determined from the liquid volume based on the fluid layer heights provided in figure 4.30 and the temperature depended liquid density $\varrho_L(\vartheta_L)$ taken from [41] as well. Thus, the total enthalpy for the entire liquid volume reads

$$dH_{e,L} = \sum_k m_{L,k}^{(II)}\,h_{e,L,k}^{(II)} - \sum_k m_{L,k}^{(I)}\,h_{e,L,k}^{(I)}\,. \tag{4.74}$$

In the following, the energy balance for the liquid is defined in more detail for the different phases; flushing, pressurization and sloshing. For flushing, the energy balance in the liquid gives

$$dQ_{L,\text{flush}} = dH_{e,L} - V_L\,dp \tag{4.75}$$

where phase change effects based on the change of the helium concentration are neglected due to the short flushing time scale. Thus, the total energy in the liquid during flushing corresponds to the inner energy. This is equivalent to a closed system.

In the next phase, the tank is pressurized with GHe. During this phase, the concentration of helium in the ullage further increases by adding more GHe. While the tank pressure increases as well, the saturation temperature of the liquid coupled to the vapor partial pressure increases only slightly. Some vapor re-condenses during the pressurization phase which has to be taken into account as well. Therefore, the energy balance for the liquid gives

$$dQ_{L,\text{press}} = dH_{e,L} - V_L\,dp - m_{\text{condens}}\left[h_{e,L,\text{condens}} + \Delta h_v\right] \tag{4.76}$$

where $m_{\text{condens}}\left[h_{e,L,\text{condens}} + \Delta h_v\right]$ is the converted vapor energy that is added to the liquid due to re-condensation during the pressurization phase, while $h_{e,L,\text{condens}}$ is the liquid enthalpy at surface level corresponding to the liquid leaving the liquid phase and Δh_v is the latent heat of vaporization that is released due to the phase change from vapor to liquid. Equation (4.69) assumes that the mass m_{condens} enters the control volume as liquid. The conversion of the vapor mass into a liquid mass requires the latent heat of vaporization. This is considered in equation (4.76). Please note that this approach is equivalent to

$$dQ_{L,\,\text{press}} = dH_{e,\,L} - V_L\,dp - m_{\text{condens}}\,h_{e,\,U,\,\text{condens}} \, .\qquad(4.77)$$

The enthalpy of the vapor mass entering the lower control volume is removed from the upper control volume as shown later in equation (4.90). Furthermore, the determination of the mass of re-condensed nitrogen m_{condens} during the pressurization phase is described afterwards in the energy balance for the ullage.

After closing the tank inlet when the required initial pressure is reached, the tank is excited. By neglecting evaporation due to the short time scale, the energy balance during the sloshing phase reads

$$dQ_{L,\,\text{slosh}} = dH_{e,\,L} - V_L\,dp - m_{\text{condens}}\left[h_{e,\,L,\,\text{condens}} + \Delta h_v\right] ,\qquad(4.78)$$

again, where $m_{\text{condens}}\left[h_{e,\,L,\,\text{condens}} + \Delta h_v\right]$ is the energy transferred into the liquid by the vapor condensation during sloshing. Analogous to the pressurization phase, $h_{e,\,L,\,\text{condens}}$ corresponds to the liquid enthalpy at surface level. During the phase change from vapor to liquid, the latent heat of vaporization Δh_v is released and therefore absorbed by the liquid. Similar to the pressurization phase, this approach is equivalent to

$$dQ_{L,\,\text{slosh}} = dH_{e,\,L} - V_L\,dp - m_{\text{condens}}\,h_{e,\,U,\,\text{condens}}\qquad(4.79)$$

where the enthalpy of the vapor mass entering the lower control volume is removed from the upper control volume and the released latent heat of vaporization is absorbed by the lower control volume. Furthermore, the determination of the condensed mass is described afterwards in the energy balance of the ullage.

The ullage is considered as open 2-species system where vapor can either enter (evaporation) or leave (condensation) via the free surface, while the pressurant gas (GHe) entering the tank must be considered as well. Furthermore, the tank is flushed with helium that is realized by injecting GHe through the gas inlet, while GN_2 in the vicinity of the free surface is flushed through the venting pipe. Thus, the energy balance of the ullage can be defined in general by

$$dQ_U = dU_{e,\,U} + \sum m_{U,\,\text{out}}\,h_{e,\,U,\,\text{out}} - \sum m_{U,\,\text{in}}\,h_{e,\,U,\,\text{in}}\qquad(4.80)$$

where the helium that is added during flushing/pressurization through the gas inlet as well as evaporation are regarded in the source term $m_{U,\,\text{in}}\,h_{e,\,U,\,\text{in}}$ and gaseous nitrogen that is vented during flushing as well as condensation are regarded in the sink term $m_{U,\text{out}}\, h_{e,U,\text{out}}$. According to equation (2.17), the inner energy of the ullage can be expressed by the enthalpy yielding

$$dU_{e,U} = dH_{e,U} - p\,dV_U - V_U\,dp\,,\tag{4.81}$$

again, where $p\,dV_U = 0$ for constant volumes. As well in the ullage, the change of the enthalpy is defined as difference between the enthalpy in state (I) and state (II), so that

$$dH_{e,U} = H_{e,U}^{(II)} - H_{e,U}^{(I)} = m_U^{(II)}\, h_{e,U}^{(II)} - m_U^{(I)}\, h_{e,U}^{(I)}\,,\tag{4.82}$$

which can be rewritten for a 2-species system composed of GHe and GN$_2$. Thus,

$$dH_{e,U} = \left[m_{U,\text{GHe}}^{(II)}\, h_{e,\text{GHe}}^{(II)} + m_{U,\text{GN2}}^{(II)}\, h_{e,\text{GN2}}^{(II)} \right] - \left[m_{U,\text{GHe}}^{(I)}\, h_{e,\text{GHe}}^{(I)} + m_{U,\text{GN2}}^{(I)}\, h_{e,\text{GN2}}^{(I)} \right]\,.\tag{4.83}$$

As shown in figure 4.30, the ullage is subdivided into k layers based on the distances between the temperature sensors. The specific enthalpy in the ullage is determined for each layer corresponding to the measured temperature at sensor position T1/T5 to T4/T8 by using [41]. The total gas mass in each layer is composed of a helium part $m_{U,\text{GHe},k}$ and a vapor part $m_{U,\text{GN2},k}$, so that

$$m_{U,k} = m_{U,\text{GHe},k} + m_{U,\text{GN2},k}\,.\tag{4.84}$$

The mass fractions of GHe and GN$_2$ are determined based on their concentrations, by the gas volume based on the liquid layer heights as shown in figure 4.30, by the temperature depended gas density taken from [41] and by the known mass fraction x as provided in table A.4 in the appendix. Thus, the helium mass in the layer k gives

$$m_{U,\text{GHe},k} = x_{\text{GHe}}\, \varrho_{U,\text{GHe},k}\left(\vartheta_{U,k}\right) V_{U,k}\tag{4.85}$$

and analogous the nitrogen vapor mass gives

$$m_{U,\text{GN2},k} = x_{\text{GN2}}\, \varrho_{U,\text{GN2},k}\left(\vartheta_{U,k}\right) V_{U,k}\,.\tag{4.86}$$

Thus, the enthalpy for the entire ullage volume consisting of two species and k layer reads

$$
\begin{aligned}
dH_{e,U} = \sum_k &\left[m_{U,\text{GHe},k}^{(II)}\, h_{e,\text{GHe},k}^{(II)} + m_{U,\text{GN2},k}^{(II)}\, h_{e,\text{GN2},k}^{(II)} \right] - \\
\sum_k &\left[m_{U,\text{GHe},k}^{(I)}\, h_{e,\text{GHe},k}^{(I)} + m_{U,\text{GN2},k}^{(I)}\, h_{e,\text{GN2},k}^{(I)} \right]\,.
\end{aligned}\tag{4.87}
$$

In the following, the energy balance for the ullage is defined in detail for the different phases; flushing, pressurization and sloshing. For the flushing phase where helium is injected through the gas inlet, while vapor is vented through the gas outlet at isobaric conditions, the energy balance in the ullage reads

$$dQ_{U,\text{flush}} = dH_{e,U} - V_U\, dp + m_{\text{GN2},f}\, h_{e,\text{GN2}} - m_{\text{GHe},f}\, h_{e,\text{GHe}} \tag{4.88}$$

where phase change effects based on the variation of the helium concentration are neglected due to the short flushing time scale. Furthermore, $m_{\text{GN2},f}\, h_{e,\text{GN2}}$ is the energy of the gaseous nitrogen that is removed from the tank during flushing and $m_{\text{GHe},f}\, h_{e,\text{GHe}}$ is the energy of the helium that is brought into the tank to replace the GN_2. The helium mass that is fed during flushing is provided in table A.4 in the appendix. Assuming that the volume fraction of the replaced gaseous nitrogen is equal to the volume fraction of the injected helium, the GN_2 mass is determined by

$$m_{\text{GN2},f} = m_{\text{GHe},f}\, \frac{\varrho_{\text{GN2}}\left(\vartheta_{\text{GN2}}\right)}{\varrho_{\text{GHe}}\left(\vartheta_{\text{GHe}}\right)}. \tag{4.89}$$

The helium density is $\varrho_{\text{GHe}} = 0.16$ kg m^{-3} according to a gas temperature of $\vartheta_{\text{GHe}} = 293$ K, while the gaseous nitrogen density is $\varrho_{\text{GN2}} = 4.61$ kg m^{-3} according to a gas temperature of $\vartheta_{\text{GN2}} = 77.4$ K slightly above the liquid surface. The enthalpy is determined accordingly, so that $h_{e,\text{GHe}} = 1527$ kJ kg^{-1} and $h_{e,\text{GN2}} = 77.25$ kJ kg^{-1}.

During pressurization, the venting line is closed, so that only GHe is injected through the gas inlet. While increasing the tank pressure with the non-condensable gaseous helium, the saturation temperature of the liquid increases only slightly. During this phase, a certain amount of vapor re-condenses to be considered as well. Thus, the energy balance in the ullage for the pressurization phase reads

$$dQ_{U,\text{press}} = dH_{e,U} - V_U\, dp - m_{\text{GHe},f+p}\, h_{e,\text{GHe}} + m_{\text{condens}}\, h_{e,U,\text{condens}} \tag{4.90}$$

where $m_{\text{GHe}}\, h_{e,\text{GHe}}$ is the energy of the helium to pressurize the tank. The helium mass $m_{\text{GHe},f+p}$ to pressurize the tank is provided in table A.4 in the appendix. The enthalpy of the helium during pressurization is $h_{e,\text{GHe}} = 1527$ kJ kg^{-1} for a gas temperature of $\vartheta_{\text{GHe}} = 293$ K. Furthermore, $m_{\text{condens}}\, h_{e,U,\text{condens}}$ is the energy leaving the ullage due to re-condensation during GHe pressurization, while $h_{e,U,\text{evap}}$ corresponds to the vapor enthalpy at surface level. The re-condensed mass is determined by integrating the according nitrogen vapor mass over all gas layers (see figure 4.30) in state (I) corresponding to the initial state after flushing and in state (II) corresponding to the final state after pressurization, so that

$$m_{\text{condens}} = m_L^{(II)} - m_L^{(I)} = \sum_k m_{L,k}^{(II)} - \sum_k m_{U,k}^{(I)} \tag{4.91}$$

where k is the index corresponding to the different gas layer. For the pressurization phase, the vapor mass in each layer is determined by

$$m_{L,k} = \varrho_{L,k}\left(\vartheta_{L,k}\right) V_{L,k}. \tag{4.92}$$

Note that for the GHe pressurization cases, the liquid temperatures are taken into account to determine the condensed mass fraction m_{condens} contrary to the self-pressurization and GN_2 pressurization cases where the phase change mass is determined based on the ullage temperatures. On the one side, the gas phase containing two species is not uniformly mixed leading to a higher uncertainty by determining the ullage mass. On the other side, the temperature gradients within the liquid particularly with increasing helium concentration are smaller than observed in the one-species systems implying smaller uncertainties particulary in the vicinity of the free surface.

During the sloshing phase, all inlets and outlets are closed. Mass can only pass the control volume boundary via the free surface in terms of phase change (condensation). Thus, the energy balance in the ullage for the sloshing phase reads

$$dQ_{U,\text{slosh}} = dH_{e,U} - V_U\,dp + m_{\text{condens}}\,h_{e,U,\text{condens}}\,. \tag{4.93}$$

Again here, $m_{\text{condens}}\,h_{e,U,\text{condens}}$ is the energy that is removed from the ullage phase due to vapor condensation during the sloshing phase. Analogous to the pressurization phase, $h_{e,U,\text{condens}}$ corresponds to the ullage enthalpy at surface level. The condensed mass is determined according to equation (4.91) and equation (4.92).

The energy balance for the flushing phase is shown in figure 4.31 as function of the ullage helium concentration after flushing $\chi_{\text{GHe},f}$. In total, the energy of the system does not change significantly, while increasing the flushing duration and therefore the ullage helium concentra-

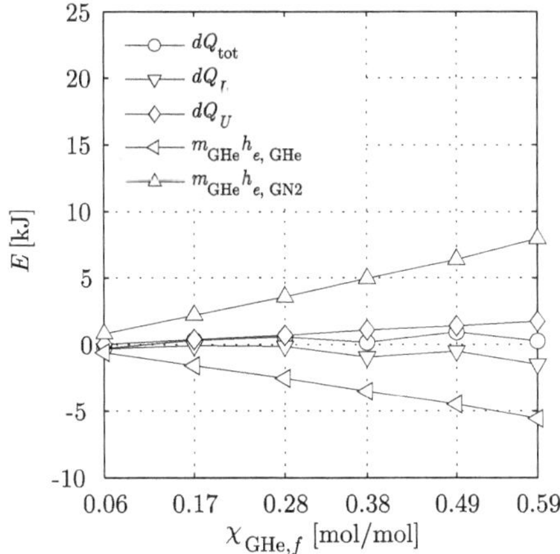

Figure 4.31: Energy balance for GHe flushing phase where GHe is fed into the system, while GN_2 is vented. The solid lines are connections between the data points to increase the readability.

tion as indicated by the ○ symbols. When the energy of the ullage increases for increasing flushing duration using warm helium, the energy of the liquid slightly decreases with increasing flushing duration. During flushing, the system is open having an inlet and an outlet, so that the pressure is approximately constant inside the tank. Helium is a non-condensable gas that does not support condensation. The liquid is less affected by the flushing than the ullage where the gas exchange occurs. The slightly decrease of the liquid energy can be explained by the slightly decrease of the surface temperature layer due to the change of the nitrogen partial pressure inside the tank during flushing.

The energy balance for the pressurization phase is shown in figure 4.32 (A) as function of the helium concentration in the ullage after flushing and pressurization $\chi_{\mathrm{GHe},f+p}$. The impact of the helium concentration is only marginal during pressurization. By trend, only a slightly decreasing change of energy is observed for dQ_L and dQ_{tot}, while the energy in the ullage slightly increases. As shown in table A.4 in the appendix, the duration of GHe pressurization t_{f+p} decreases to reach the initial pressure of $p_0 = 140$ kPa for increasing flushing duration t_f. Since m_{GHe} slightly decreases, the energy of the pressurant gas $m_{\mathrm{GHe}}\,h_{e,\mathrm{GHe}}$ slightly decreases as well for increasing $\chi_{\mathrm{GHe},f+p}$. Contrary to the $\mathrm{GN_2}$ case, the nearly constant energy of the liquid can be explained by the fact that only the helium partial pressure changes during the pressurization but the $\mathrm{GN_2}$ partial pressure remains approximately constant, so that the liquid is only marginally affected by the GHe pressurization.

The energy balance for the GHe pressurized system during sloshing phase is provided in fig-

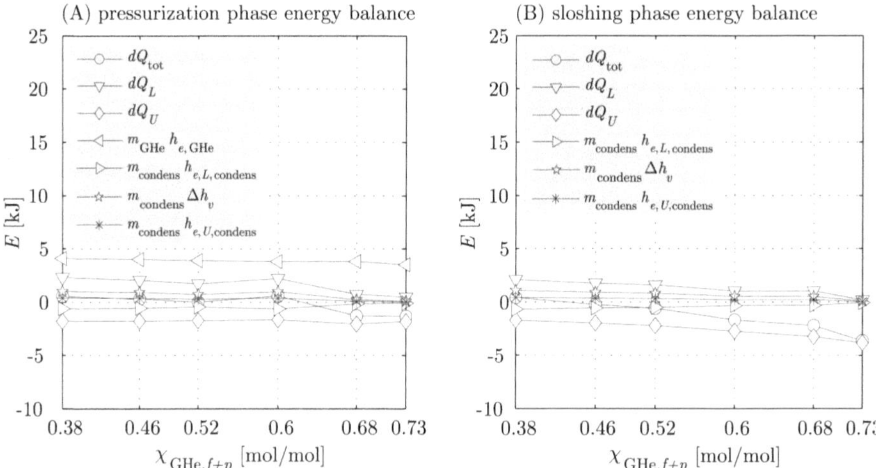

Figure 4.32: Energy balance of $\mathrm{GN_2}$ pressurized systems for the pressurization phase (A) and for the sloshing phase (B). The solid lines connect the data points to increase the readability.

ure 4.32 (B) as function of the helium concentration in the ullage after flushing and pressurization. For smaller GHe concentrations, the process is driven by condensation initiated by mixing effects in the thermal boundary layer under the free liquid surface where the surface temperature and therefore the saturation temperature decreases. The mixing depends strongly on the helium concentration as shown in figure 4.29 (A). For increasing helium concentrations, the liquid appears less stratified with a decreased temperature gradient under the free liquid surface. The surface temperature is set by means of the GN_2 partial pressure $p_{\text{GN2}} = p - p_{\text{GHe}}$ and therefore by the helium concentration. Furthermore, the change of liquid energy decreases as indicated by the ∇ symbols. The change of energy in the ullage dQ_U decreases as well driven by cooling-down the ullage gas provoked by the sloshing liquid. The energy coupled to the phase change on the liquid side ($\triangleright$ symbols) slightly increases while the energy coupled to the phase change on the ullage side ($*$ symbols) slightly decreases for increasing helium concentration. From a certain helium concentration, condensation disappears for increasing $\chi_{\text{GHe},f+p}$. In fact, condensation stops for ullage helium concentrations larger than approximately $\chi_{\text{GHe},f+p} \geq 0.73 \, \text{mol}\,\text{mol}^{-1}$.

Likewise as observed for the one-species experiments, the pressure drop occurring under the impact of sloshing for the two-species experiments is superimposed by two different effects; the ullage cooling-down as well as condensation effects caused by the decrease of the surface temperature. This is shown in figure 4.33 where the full $\blacktriangleleft$ symbols correspond to the total pressure drop measured by the pressure sensor within the tank lid indicated by $dp_{\text{press drop}}$. By knowing the mean gas temperature and the mean mass distribution from the ullage gas densities $\varrho_{U,k}(\vartheta_{U,k})$, the pressure drop fraction caused by the temperature decrease of the ullage gas is determined by means of the ideal gas law considering the two-species system, so that

$$dp_{\text{ullage temp}} = \frac{1}{V_U} m_{U,\text{GHe}} R_{s,\text{GHe}} d\bar{\vartheta}_U + \frac{1}{V_U} m_{U,\text{GN2}} R_{s,\text{GN2}} d\bar{\vartheta}_U \tag{4.94}$$

with

$$m_{U,\text{GHe}} = \sum_k \chi_{\text{GHe},f+p} \, \varrho_{U,\text{GHe},k}(\vartheta_{U,k}) \, V_{U,k} \tag{4.95}$$

and

$$m_{U,\text{GN2}} = \sum_k \chi_{\text{GN2},f+p} \, \varrho_{U,\text{GN2},k}(\vartheta_{U,k}) \, V_{U,k} \, . \tag{4.96}$$

Moreover, $R_{s,\text{GHe}} = 2077 \, \text{J}\,\text{kg}^{-1}\,\text{K}^{-1}$ is the specific gas constant for GHe, while the specific gas constant for GN_2 gives $R_{s,\text{GN2}} = 296.8 \, \text{J}\,\text{kg}^{-1}\,\text{K}^{-1}$ as introduced in table 3.3. Furthermore, the mean temperature of the ullage is determined by

$$\bar{\vartheta}_U = \frac{1}{m_{U,\text{GHe}}} \sum_k \vartheta_{U,k} \, \chi_{\text{GHe},f+p} \, \varrho_{U,\text{GHe},k} \, V_{U,k} = \frac{1}{m_{U,\text{GN2}}} \sum_k \vartheta_{U,k} \, \chi_{\text{GN2},f+p} \, \varrho_{U,\text{GN2},k} \, V_{U,k} \, . \tag{4.97}$$

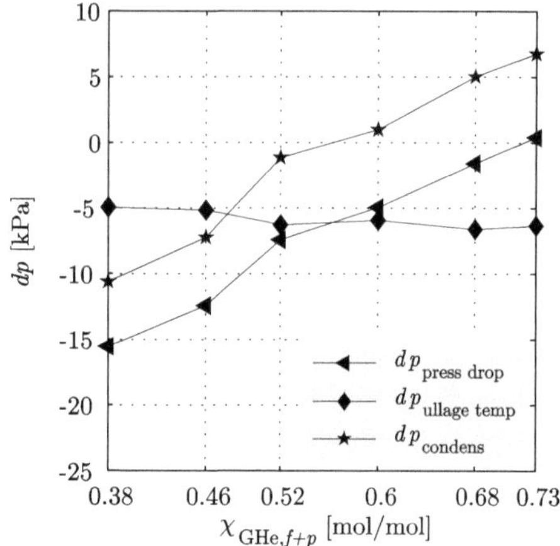

Figure 4.33: Composition of the pressure drop for the GHe pressurized systems showing the pressure change in the ullage. The solid lines connect the data points to increase the readability.

As the helium concentration increases, the temperature of the ullage remains approximately constant as indicated by the full ♦ symbols. Furthermore, the pressure drop caused by condensation is determined by

$$dp_{\text{condens}} = dp_{\text{press drop}} - dp_{\text{ullage temp}} \, , \tag{4.98}$$

which is indicated by the full ★ symbols. For $\chi_{\text{GHe},f+p} > 0.6 \, \text{mol}\,\text{mol}^{-1}$, the pressure drop caused by condensation becomes positive. In fact, this is can be considered as the change to a pressure rise indicating evaporation. Indeed, the measured pressure drop is still negative, but figure 4.33 indicates that the process is driven by the temperature decrease of the ullage gas, while condensation is disabled by the appearance of the non-condensable gas. This threshold concentration is slightly smaller than the value provided based on figure 4.32.

4.3.5 Upscaling of the Pressure Drop Data

A major objective of this work includes to implement an appropriate upscaling concept to validate the possibility of extrapolating the actual data to gain information for the full size application. Considering a one-species system fed with liquid nitrogen (LN$_2$), the initial pressure is linearly increased stepwise to test the capability for an extrapolation purpose and therefore to predict the impact of the pressure drop within the hydrogen tank compartment on a cryogenic upper stage under the impact of sloshing. For high pressure predictions according to the full size application, the actual data as well as previous results from the literature [48, 39, 26] are merged to gain information about the quality of the actual scaling law for similar conditions regarding a tank that is 50% full. The impact of the pressure drop is characterized by two scaled numbers; the pressure gradient $|\partial p^*/\partial t^*|$ and the pressure drop magnitude $\Delta p^*_{\mathrm{mag}}$ that are both determined from the actual pressure developments shown in figures 4.6, 4.15 and 4.26. An illustration of the pressure gradient as well as the pressure drop magnitude is provided in figure 4.34 showing a typical pressure development during pressurization and sloshing phase (e.g. self-pressurization with $p^*_0 = 1.32$).

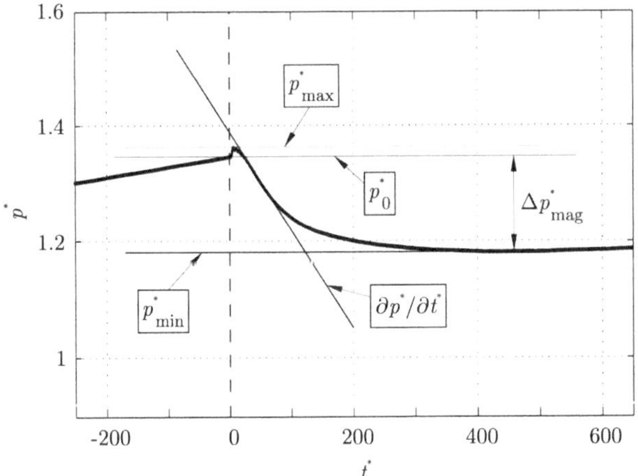

Figure 4.34: Pressure gradient $|\partial p^*/\partial t^*|$ and pressure drop magnitude $\Delta p^*_{\mathrm{mag}}$. This illustration shows a typical pressure drop for a self-pressurized system with $p^*_0 = 1.32$ (thick solid line). The dashed line indicates the initiation of the excitation, while the thin solid lines correspond to the level of the minimum pressure p^*_{min}, the maximum pressure p^*_{max} and the initial pressure p^*_0. The inclined line corresponds to the highest pressure gradient after passing the maximum pressure p^*_{max}.

Pressure Gradient

For the layout of the helium budget concerning future cryogenic upper stages, it is of high importance to estimate the dynamic impact of the pressure drop characterized by the pressure gradients during the excited phase, e.g. the ascent phase. The pressure gradient $|\partial p^*/\partial t^*|$ gives the pressure loss per time due to sloshing providing important information about the time response of the system that is exposed to lateral disturbances. Determined from the experimental pressure data, the pressure gradient is defined as

$$\left|\frac{\partial p^*}{\partial t^*}\right| = \left.\frac{p^*_{i+1} - p^*_i}{dt^*_{\min}}\right|_{\max} \tag{4.99}$$

where p^*_i is the tank pressure at t^*_i, while p^*_{i+1} is the tank pressure at $t^*_i + dt^*_{\min}$ between $p^*_{\max} \leq p^* \leq p^*_{\min}$. The minimum time step $dt^*_{\min} = 0.014$ corresponds to the frame rate for pressure data acquisition of 100 Hz. However, the pressure gradient provides important information about the time response of the system that is exposed to lateral disturbances.

For the actual experiments, the pressure gradients are shown in figure 4.35 as function of the according initial pressures p^*_0. Hereby, the semi logarithmic axis scale is appropriate to present the actual data combined with results from the literature [48, 39, 26] and the full size application ranging over several orders of magnitude. The one-species experiments conducted for this work are marked by open symbols. The bold solid lines represent linear fits[5] to allow the linear extrapolation of the actual data. The equations for these fits are provided in figure 4.35. The dashed lines indicate an error of approximately $\pm 15\%$ that is based on variations of the actual data including the measurement inaccuracy and the uncertainty due to the time derivative for determining the pressure gradient. Obviously, the pressure gradients are higher for the external GN_2 pressurization experiments than for the self-pressurization experiments by approximately a factor of 2. This can be traced back to differences in the thermal boundary layer depth δ_t below the free liquid surface. According to the pressurization time, the boundary layer is more developed for self-pressurization experiments as for the external pressurized experiments. Furthermore, the liquid in the bulk in a self-pressurized tank is less subcooled with respect to $\vartheta_{\mathrm{sat},0}$ as shown in figure B.4 in the appendix. Hence, the thermal gradients under the free liquid surface are smaller leading to pressure drops of smaller extent.

The gray symbols in figure 4.35 correspond to the two-species experiments that are flushed and pressurized with helium (GHe). Here, the gray circle symbol accords to the smallest helium concentration in the ullage ($\chi_{\mathrm{GHe},f+p} = 0.38$). For increasing χ_{GHe} as indicated by the gray triangle ($\chi_{\mathrm{GHe},f+p} = 0.52$) and gray diamond symbols ($\chi_{\mathrm{GHe},f+p} = 0.73$), the magnitude of the pressure gradient decreases until the pressure drop entirely disappears for $\chi_{\mathrm{GHe}} \gtrsim 0.73$ where the tank pressure further increases although the tank is excited. The pressure gradients shown in figure 4.35 correspond only to pressure drops (negative gradients).

---

[5]Note that the exponential shape of these lines is due to the logarithmic scaling of the y-axis

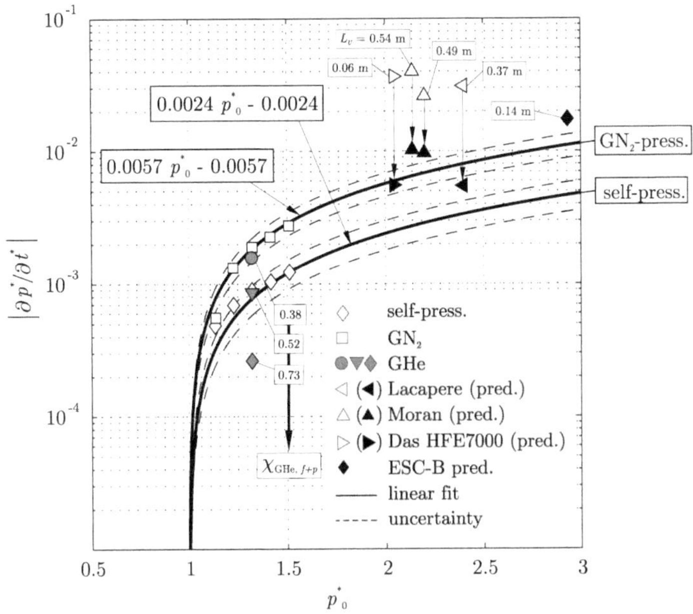

Figure 4.35: Magnitude of the pressure gradient $|\partial p^*/\partial t^*|$. The open symbols represent the actual one-species data, while the gray symbols correspond to the actual helium pressurized system. The solid lines are fits to the actual data, while the dashed lines give a uncertainty of $\pm 15\%$. Furthermore, predictions based on data from the literature - LACAPERE *et al.* [39], MORAN *et al.* [48], DAS & HOPFINGER [26] - and the ESC-B upper stage are indicated by full symbols.

The results based on the data from the literature are derived from experiments performed with liquid hydrogen (LH$_2$) [48], liquid oxygen (LOX) [39] and HFE7000 [26] under higher initial pressures as regarded here. By applying the introduced scaling, these results are correlated to the predicted pressure gradient development presented in figure 4.35. The data extrapolation is indicated by the bold solid lines. For the experiments presented in [48], [39] and [26], the excitation is in the vicinity of the corresponding first resonance frequency. Thus, the frequency ratio is in the order of $\eta_{11} \approx 1$ provoking large amplitude sloshing waves which is in contrast to $\eta_{11} = 0.78$ as chosen here. The $\triangleleft$ symbol indicates the result provided by LACAPERE *et al.* [39] performing sloshing tests in a cylindrical $R = 0.095$ m tank having a height of $H_{tot} = 0.8$ m by using LOX. The ullage volume of 55% of the total tank volume and the characteristic length of $L_U = 0.37$ m are similar to the condition as considered for this work. The experiment is pressurized with gaseous oxygen (GOX) up to an initial pressure of $p_0 = 250$ kPa within 100 s corresponding to a scaled initial pressure of $p_0^* = 2.4$. The experiment is excited with a frequency of $f = 2.1$ Hz and an amplitude of $y_A = 3$ mm.

The $\triangle$ symbols indicate the experiments performed by MORAN et al. [48] using a spherical $R = 0.75$ m tank that is partly filled with LH_2. The total volume of the tank is $V_{tot} = 1.756$ m^3. The number attached to the symbols is the characteristic length scale L_U corresponding to different fill level and therefore to different ullage volumes in a spherical tank. In this context, $L_U = 0.49$ m (case 243) corresponds to a fill level of $H = 0.763$ m and a ullage volume to total volume ratio of $V_U/V_{tot} = 49$ %, while $L_U = 0.54$ m (case 887) corresponds to $H = 0.713$ m and $V_U/V_{tot} = 54$ %. Also here, the ullage volume of the considered experiments and the characteristic length scale are similar to the values considered in this work to ensure comparability. Using gaseous hydrogen (GH$_2$) as pressurant gas, the initial pressure is $p_0 = 239$ kPa ramped up within 20 s corresponding to a scaled initial pressure of $p_0^* = 2.14$ for the $L_U = 0.49$ m experiment, while for the $L_U = 0.54$ m experiment, the initial pressure is $p_0 = 246$ kPa ramped up within 15 s corresponding to a scaled initial pressure of $p_0^* = 2.20$. Both experiments are excited with a frequency of $f = 0.74$ Hz and an amplitude of $y_A = 38.1$ mm.

The result based on data from DAS & HOPFINGER [26] is indicated by the $\triangleright$ symbol. They used a cylindrical $R = 0.05$ m tank that is filled with engineering fluid HFE7000. The total volume of the tank is $V_{tot} = 9.03\text{E-}4$ m^3. The fill level is $H = 0.055$ m and the characteristic length scale gives $L_U = 0.06$ m corresponding to a total volume to ullage volume ratio of $V_U/V_{tot} = 0.52$ %. The characteristic length is a factor 6 smaller than the L_U considered in this work. The initial pressure for the HFE7000 experiment is $p_0 = 229$ kPa ramped up by HFE7000 vapor injection in within 79 s including a hold period of 20 to 30 s. This initial pressure corresponds to a scaled initial pressure of $p_0^* = 2.04$. The experiment is excited with a frequency of $f = 2.925$ Hz and an amplitude of $y_A = 1.135$ mm.

Regarding the higher excitation frequencies and therefore higher wave amplitudes, the pressure gradient values ($\triangle$, $\triangleleft$ and $\triangleright$) appear higher than they would have been predicted by the linear extrapolations of the actual external pressurization data as indicated by the upper solid line. As introduced in chapter 2, DAS & HOPFINGER [26] found that the sloshing wave amplitude has a significant influence on the development of the pressure gradient. Therefore, the literature data [48, 39, 26] is recalculated with respect to the excitation used in this work as described in the following.

According to [26], the effective diffusion coefficient can be considered as a measure of the mixing in the thermal boundary layer that is related to the excitation of the tank. This is defined in equation (2.93) yielding

$$D_e = -u_v \frac{\delta_t}{\mathbf{R\,Ja}} \tag{4.100}$$

or as function of the nondimensional diffusion coefficient $D_e'^*$ to express the similarity of the excitation and therefore the sloshing wave amplitude in different sized tanks. Thus,

$$D_e = D_e'^* \left(g\,R^3\right)^{1/2} + D_T. \tag{4.101}$$

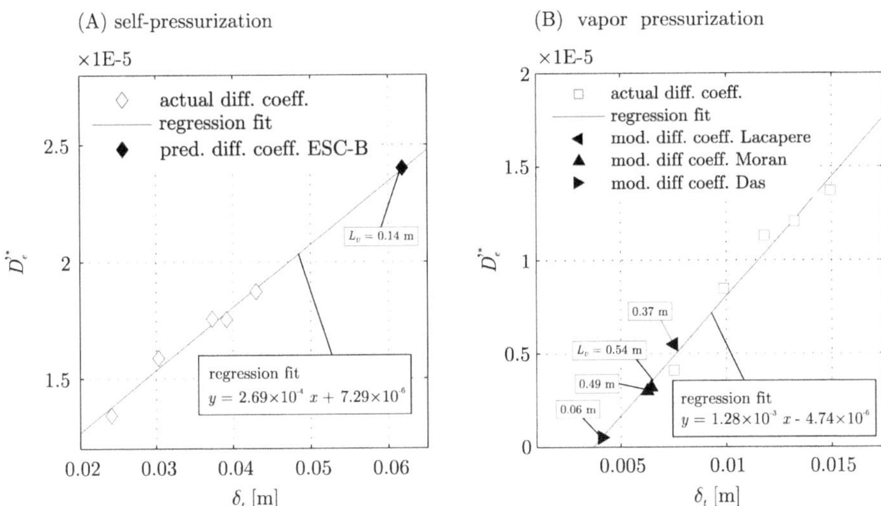

Figure 4.36: The nondimensional diffusion coefficient $D_e^{'*}$ for self-pressurization (A) and for vapor pressurization (B). The solid lines correspond to linear regressions based on the current data with the coefficient of determination $R^2 = 0.92$ in case of (A) and $R^2 = 0.89$ in case of (B). The open symbols correspond to the actual data, while the full symbols indicate the modified diffusion coefficients of previous experiments [39, 48, 26] and the ESC-B prediction by correlating the known thermal boundary layer depth δ_t with the development of $D_e^{'*}$.

Beside the excitation, the dimensionless diffusion coefficient $D_e^{'*}$ depends also on the initial pressure and therefore on the pressurization history of the tank that is basically expressed by the thermal boundary layer depth δ_t. The difference between self pressurization and external pressurization is shown in figure 4.36 where $D_e^{'*}$ is plotted as function of the thermal boundary layer depth δ_t. The data is summarized in table A.45, A.46 and A.47 in the appendix. The solid lines correspond to linear regressions to fit this data.

According to equation (2.84), the pressure drop is composed of the pressure decrease due to vapor condensation and of the pressure decrease due to cooling-down the gas phase. With respect to equations (4.100) and (4.101), equation (2.84) can be rewritten

$$\frac{\partial p}{\partial t} = -\frac{1}{L_U} \left[D_e^{'*} \left(g\, R^3 \right)^{1/2} + D_T \right] \frac{R\, \mathrm{Ja}}{\delta_t} p_0 + \frac{p_0}{\vartheta_{U,0}} \frac{\partial \vartheta_U}{\partial t} \tag{4.102}$$

where $p_0 = \varrho_{U,0}\, R_s\, \vartheta_{U,0}$ gives the initial pressure. As already introduced, the temperature gradient in the liquid is expressed by the thermal boundary layer depth δ_t. To regard similar conditions such as the lateral sloshing motion of the experiments performed for this work, $D_e^{'*}$ is determined by the correlations provided in figure 4.36 with respect to the original δ_t found in [39] for LOX, in [48] for LH_2 and in [26] for HFE7000. In detail, this is explained in the following.

For the LOX experiment conducted by LACAPERE *et al.* [39], where chaotic sloshing near resonance was observed, the pressure gradient is $\partial p / \partial t = -8300$ Pa s^{-1} taken from [26] for a tank that is pressurized with vapor (GOX) up to an initial pressure of $p_0 = 250$ kPa. Based on equation (2.91) the thermal boundary layer depth was recalculated by [26] taking into account data from [39] to find $\delta_t = 0.0075$ m. According to figure 4.36 (B), a diffusion coefficient prediction is chosen based on the actual experimental data for tanks that are vapor pressurized and sloshed in the lateral regime based on an excitation of $\eta_{11} = 0.78$. Thus, the dimensionless diffusion coefficient for the same boundary layer depth δ_t but an excitation of $\eta_{11} = 0.78$ gives $D_e'^* = 5.5\text{E}-6$. The modified pressure gradient is indicated by the ◄ symbol in figure 4.35. The result slightly underpredicts the extrapolation for external vapor pressurization corresponding to the upper solid line. Tabular data to determine this data point is provided in table A.48 and A.49 in the appendix.

This method is also applied to recalculate MORAN's data [48] who used liquid hydrogen in a $R = 0.748$ m tank that is excited close to its first natural frequency where $\eta_{11} = 0.95$. For $L_U = 0.49$ m, they observed a pressure gradient of $\partial p / \partial t = -2344$ Pa s^{-1} and for $L_U = 0.54$ m, they observed a pressure gradient of $\partial p / \partial t = -3585$ Pa s^{-1}. Analogous to the previous paragraph, the dimensionless diffusion coefficient assuming an excitation in the lateral regime in the order of $\eta_{11} = 0.78$ for an initial pressure of $p_0 \approx 242$ kPa is predicted based on a thermal boundary layer depth of $\delta_t = 0.0063$ m. Thus, $D_e'^* = 3.0\text{E}-6$ for an ullage volume to surface area ratio of $L_U = 0.54$ m and $D_e'^* = 3.2\text{E}-6$ for $L_U = 0.49$ m. Result are depicted by ▲ symbols that are as well in good agreement to the linear prediction for external pressurization indicated by the upper solid line. Tabular data to determine this data point is provided in table A.50 and A.51 in the appendix.

Furthermore, the ► symbol corresponds to the recalculation of HFE7000 data published by DAS & HOPFINGER [26]. For $L_U = 0.06$ m, they observed a pressure gradient of $\partial p / \partial t = -4100$ Pa s^{-1}. Based on equation (2.91), the thermal boundary layer depth is re-determined to be $\delta_t = 0.0041$ m. Assuming the lateral sloshing regime for an excitation in the order of $\eta_{11} = 0.78$, the dimensionless diffusion coefficient gives $D_e'^* = 5.2\text{E}-7$. The result is in good agreement to the linear prediction for the vapor pressurized system. Tabular data to determine this point is provided in table A.54 and A.55 in the appendix.

The ◆ symbol in figure 4.35 represents the prediction of the characteristic pressure drop that might occur in the hydrogen tank compartment on the ESC-B upper stage of Ariane 5 assuming a fill level of 95% during the ascent phase where the tank may be laterally excited due to different flight maneuvers. In fact, the excitation of the tank must be considered as 6-axis problem including lateral accelerations in x, y, and z direction as well as the according angular accelerations corresponding to different rolling maneuvers. In general, the rolling maneuvers are executed slowly and it is assumed that the propellant sloshing is mainly driven by lateral accelerations in x and y direction during the ascent phase. It is assumed that this is sufficient to excite the liquid in the first asymmetric sloshing (lateral) mode. The predictions for the pressure drop inside the ESC-B upper stage hydrogen tank compartment are determined based on equivalent conditions as found in MORAN's experiments performed with liquid hydrogen [48]. For the ESC-B scenario, a tank radius of $R = 2.7$ m is assumed, while the first natural frequency of the tank regarding a fill level of 95% is estimated in the order of $f_{11} \approx 0.8$ Hz. Taking into account a frequency ratio of $\eta_{11} = 0.78$ as defined for the current work, the excitation frequency of the tank gives approximately $f \approx 0.6$ Hz assuming the lateral sloshing mode. Due to the long duration holding period between propellant loading and lift-off ($t_{\text{hold}} \approx 3600$ s), the tank is assumed to be self-pressurized[6] up to an initial pressure of $p_0^* = 2.93$ corresponding to $p_0 = 330$ kPa. Therefore, the nondimensional diffusion coefficient of $D_e'^* = 2.5\text{E}{-}5$ is determined based on the correlation provided in figure 4.36 (A) assuming a thermal boundary layer depth of $\delta_t = 0.062$ m. For a fill level of 95%, the free liquid surface is located in the upper spherical dome of the tank, so that the ullage volume to the surface ratio gives $L_U = 0.14$ m. For this configuration the prediction of the pressure gradient is $\partial p^*/\partial t^* = -0.017$ or in dimensional notation $\partial p/\partial t = -1565$ Pa s^{-1}. The result shows some variation from the prediction overestimating the pressure gradient for self-pressurization. This can be explained by the small ullage volume with respect to the liquid surface area where the impact of the pressure drop is higher. Furthermore, it is assumed that the pressure gradient can be reduced by a large extent for increasing helium concentration in the ullage as observed in the actual experiments.

Magnitude of the Pressure Drop

The magnitude of the pressure drop is the critical worst case value which might occur for a given excitation assuming steady state sloshing conditions. The magnitude of the pressure drop represents the difference between the initial pressure p_0^* reached after pressurization and its corresponding pressure minimum p_{min}^* reached during the sloshing phase. In nondimensional form, the magnitude of the pressure drop is defined as

$$\Delta p_{\text{mag}}^* = p_0^* - p_{\text{min}}^* \,. \tag{4.103}$$

For the actual experiments, the magnitude of the pressure drop is provided in figure 4.37 as function of the initial pressure and plotted on a semi-logarithmic scale. This is appropriate to adequately combine the actual data with results based on data from the literature [48, 39, 26] and the full size application ranging over several orders of magnitude. The one-species experiments conducted for this work are marked by the open symbols, while the two-species experiments are indicated by gray symbols. Literature data [48, 39, 26] and the ESC-B pre-

---

[6]Helium pressurization right before lift-off is not taken into account in order to be conservative.

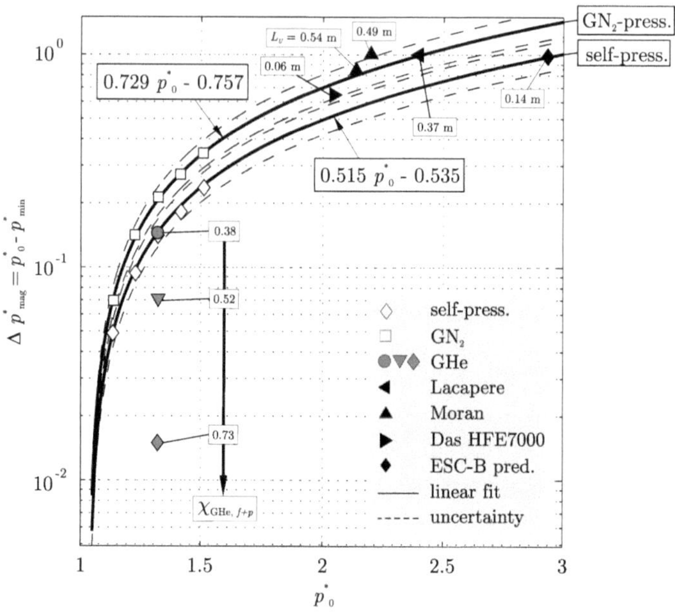

Figure 4.37: Pressure drop magnitude $\Delta p_{\mathrm{mag}}^*$. The open symbols represent the actual one-species data, while the gray symbols correspond to the helium pressurized systems. The solid lines are fits to the actual data, while the dashed lines give a uncertainty of $\pm 15\%$. Furthermore, data from the literature - LACAPERE *et al.* [39], MORAN *et al.* [48], DAS & HOPFINGER [26] - and the ESC-B are indicated by full symbols.

diction are indicated by full symbols. The solid lines correspond to linear extrapolations of the actual data, while the dashed lines indicate an inaccuracy of approximately $\pm 15\%$. The equations for the linear regression are provided in the figure.

The magnitude of the pressure is higher for the externally pressurized experiments with respect to the data conducted by self-pressurization experiments. This can be explained by differences of the thermal boundary layer depth δ_t below the free liquid surface as considered before. For self-pressurization experiments, this boundary layer is larger than for external pressurization experiments. Hence, the temperature gradients in the boundary layer are smaller for self-pressurized experiments and therefore the magnitude of the pressure drop as well.

The gray symbols in figure 4.37 indicate the two-species experiments carried out with helium flushing and pressurization. While the gray circle symbol accords to the smallest helium concentration in the ullage ($\chi_{\mathrm{GHe},f+p} = 0.38$), the magnitude of the pressure drop significantly decreases for increasing helium concentrations indicated by the gray triangle symbol. This data point represents a helium concentration of $\chi_{\mathrm{GHe},f+p} = 0.52$. The gray diamond symbol represents a helium concentration of $\chi_{\mathrm{GHe},f+p} = 0.73$. The GN_2 partial pressure decreases for increasing helium concentrations. This is a result of the decreasing surface temperature gradients in the liquid. Exceeding $\chi_{\mathrm{GHe}} = 0.73$, the pressure drop even disappears, so that the tank pressure further increases during the sloshing phase. As shown in figure 4.25 and 4.29, the temperature gradients under the free liquid surface are only marginally developed (approximately constant temperature) for high helium concentrations. Thus, liquid mixing cannot occur so that the surface temperature (saturation temperature) cannot fall and the pressure does not drop.

The ◄ symbol corresponds to the data point provided by LACAPERE *et al.* [39] performing sloshing tests in a $R = 0.095$ m cylindrical tank filled with liquid oxygen (LOX). Their experiment was based on an ullage volume to surface area ratio of $L_U = 0.37$ m corresponding to an ullage volume of approximately 50% of the total tank volume. The scaled initial pressure of this test is set to $p_0^* = 2.40$. Furthermore, this one-species experiment is externally pressurized using gaseous oxygen. The magnitude of the pressure drop $\Delta p_{\mathrm{mag}}^* = 1.0$ taken from [39] is in good agreement with the linear extrapolation for GN_2 pressurized systems.

The ▲ symbols correspond to one-species experiments performed by MORAN *et al.* [48] using liquid hydrogen (LH_2) in a $R = 0.75$ m spherical tank. Two different ullage volume to surface area ratios are considered; $L_U = 0.49$ m and 0.54 m respectively. The scaled initial pressure is $p_0^* = 2.14$ and $p_0^* = 2.20$ respectively. Also the hydrogen data is in quite good agreement to the predicted development indicated by the upper solid line. It seems that the pressure drop gains slightly in magnitude for decreasing L_U with $\Delta p_{\mathrm{mag}}^* = 0.84$ for $L_U = 0.54$ m and $\Delta p_{\mathrm{mag}}^* = 1.0$ for $L_U = 0.49$ m.

The data point performed by DAS & HOPFINGER is indicated by the ► symbol. They used a cylindrical tank filled with HFE7000 for $L_U = 0.06$ m. They found a pressure drop magnitude of $\Delta p_{\mathrm{mag}}^* = 0.65$ showing good agreement to the prediction for vapor pressurized tanks.

Other than in the previous section, the pressure drop magnitude data corresponding to higher excitation levels ($\eta_{11} \approx 1$) does not being recalculated to fit to the actual data considering a smaller excitation of the tank ($\eta_{11} = 0.78$). This implies that the pressure drop magnitude basically depends on the pressurization history and therefore on δ_t and only for a much smaller extent to the level of excitation.

The ♦ symbol in figure 4.37 corresponds to the prediction of the maximum pressure drop in the ESC-B hydrogen tank compartment taking into account a continuous excitation of $\eta_{11} = 0.78$. Thus, by assuming an scaled initial pressure of $p_0^* = 2.93$ corresponding to $p_0 = 330$ kPa, a maximum pressure drop of $\Delta p_{\mathrm{mag}}^* = 0.98$ is predicted that corresponds to $\Delta p_{\mathrm{mag}} = 110$ kPa. As well here, the application of GHe would reduce the magnitude of the pressure drop by a certain extent depending on the helium concentration in the ullage.

Chapter 5

CONCLUSIONS

In this work, the results of several cryogenic experiments are presented in order to characterize the dynamic propellant condition inside a rocket tank. Particularly during the ascent phase where the launcher is highly impacted by the atmospheric conditions, the liquid is disturbed due to several flight maneuvers such as rolling and pitching. As well, during upper stage separation the propellant is exposed to a certain lateral excitation that may cause high amplitude sloshing. The sloshing tests performed for this work enhance the understanding of the effects connected to the periodic motion of cryogenic liquids. This includes the determination of the viscous damping characteristics. Such information is of importance for the design of the upper stage attitude control system where the damping defines the transient response behavior (decay). In the main part, this work deals with the characteristic pressure drop effect that might occur under certain circumstances in the hydrogen tank compartment of the planned ESC-B upper stage as future part of the European space launcher Ariane 5. The pressure drop phenomenon magnifies the stage helium budget by a large extent to compensate this effect taking into account conservative assumptions. Therefore, different laboratory size experiments are conducted with liquid nitrogen (LN_2) as substitute for liquid hydrogen (LH_2) to gain information about the nature of this effect. This includes different pressurization methods such as self-pressurization, GN_2 pressurization, and helium pressurization in accordance to actual upper stage tank conditions. The extrapolation of the actual results allows the correlation to previous data from the literature [48, 39, 26] showing a good agreement. However, this chapter also provides the limits of this study with regards to the upscaling to predict the pressure drop of the full size application containing LH_2.

5.1 Discussion of the Damping Results

Damping experiments are conducted for a frequency ratio of $\eta = 0.78$ and an amplitude ratio of $y_A/R = 0.07$. This is sufficient to excite the liquid in the first asymmetric (lateral) sloshing mode. Under these conditions, the fill level is varied between $0.5 \leq H/R \leq 2.5$. Despite the fact that the system is highly sensible to the initial condition, the cryogenic results are in good agreement to the theoretical prediction firstly developed by MILES [46] as summarized in equation (4.8) in chapter 4. This theoretical prediction is confirmed based on experimental work by MIKISHEV *et al.* [45] and STEPHENS *et al.* [55]. In the latter case, it was found that damping can be expressed by the GALILEI number, a characteristic number that is defined as the ratio of the gravitational and the viscous forces. With respect to the GALILEI number, the damping characteristics basically depend on the tank size, the liquid kinematic viscosity and the gravitational acceleration acting on the liquid. In cylindrical tanks, the damping increases for a higher extent when the fill level undercuts a certain gauge that is typically in the range of $H/R \leq 1$ [6, 45, 55]. This implies that the tank bottom and its shape significantly gain in influence on the fluid dynamics for small fill level. Most commonly, the complex geometry of upper stage tanks varies from simple academic cylindrical shapes as studied in this work and in the literature [45, 55]. An example of the application is provided in figure 5.1 showing the planned ESC-B upper stage including the LH_2 tank, the LOX tank and the helium reservoirs. The head and bottom sections of the LH_2 tank correspond to more or less spherical half domes in convex or concave configurations providing strong influences on the dynamics of the propellant. The hatched area indicate a fill level of approximately 95% that is supposed to be the propellant gauge during the ascent phase implying that the free liquid surface is impacted by the upper tank dome only. Particularly this fact may represent the highest uncertainty to predict the liquid damping behavior in a full size upper stage tank.

The actual cryogenic damping data is in good agreement to the theoretical predictions provided previously by MIKISHEV *et al.* [45] and STEPHENS *et al.* [55]. For high fill level $H/R > 1$, a damping coefficient of $K_{\text{deep}} = 0.83$ is found confirming as well the qualification of the theory

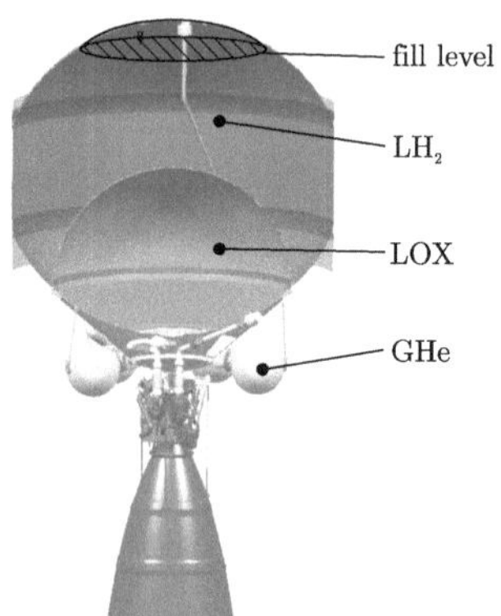

Figure 5.1: The planned restartable cryogenic upper stage ESC-B of the European space launcher Ariane 5 including the tank compartments for LH_2, LOX, and GHe [17].

for liquid nitrogen. Furthermore, an experimental relation found by MIKISHEV *et al.* [45] to consider the enhanced damping properties of the spherical bottom geometry is experimentally confirmed for cryogenic liquids (LN$_2$) for fill level between $0.5 \leq H/R \leq 2.5$. In particular for cylindrical tanks, the GALILEI number is a good approach to determine the expected damping characteristics in cylindrical tanks. However, the complex geometry of current upper stage tanks impede precise predictions with respect to viscous damping. Regarding complex geometries such as shown in figure 5.1, the geometry factor K has to be determined by individual experiments or respective CFD simulations for adequate damping predictions.

Commonly, liquid hydrogen (LH$_2$) is utilized as rocket fuel for cryogenic upper stage engines. The damping prediction for a cylindrical full size tank with spherical bottom geometry containing liquid hydrogen is presented in order to satisfy equation (4.8). For a similar tank geometry and a fill level $H/R \geq 1$, the damping coefficient for deep tanks is assumed to be $K_{\mathrm{deep}} = 0.83$ as well. The prediction agrees with the result achieved with liquid nitrogen since their kinematic viscosities almost coincide. By increasing the tank size, the damping decreases for some extent. For large reservoirs, the influence of the STOKES boundary layer between the liquid and the wall decreases in impact implying that the liquid motion is dominated by the viscous damping of the liquid only. Here, a damping ratio in the order of $\gamma \sim 10^{-4}$ is found.

5.2 Discussion of the Pressure Drop Results

In order to demonstrate the characteristic pressure drop effect, which might occur on cryogenic upper stages under certain circumstances, cryogenic sloshing experiments in laboratory size are conducted under the variation of the initial conditions. This includes three different pressurization methods; self-pressurization, GN$_2$ pressurization and helium pressurization. These pressurization methods represent different procedures that may come into consideration to configure the future upper stage. The laboratory size facility for cryogenic sloshing tests used for the experiments described in this work allows initial pressures up to $p_0 = 160$ kPa, which is consistent with a scaled initial pressure of $p_0^* = 1.51$ in the actual configuration. This includes an ullage volume of approximately 50% ($L_U = 0.36$ m). Typical tank pressures in upper stage tanks range between $2.2 \leq p_0^* \leq 3.1$, while the ullage volume in the propellant tank during the ascent phase is in the order of 5% ($L_U \approx 0.1$ m). As pointed out in the previous section, the wave motion of the liquid strongly depends on the tank shape. For this work, the liquid is excited in the first asymmetric sloshing mode with a frequency ratio of $\eta = 0.78$ and an amplitude ratio of $y_A/R = 0.069$. This leads to a surface deflection with respect to the horizontal of approximately 14.7 degrees assuming a flat surface. Similar conditions are defined as worst case for the upper stage tanks during the ascent phase whereas the resulting wave amplitude is assumed to be smaller for some extent since the tank geometry corresponding to a fill level of 95% may allow higher damping characteristics due to the spherical upper dome geometry.

Self-pressurization experiments are driven by means of the heat flowing into the tank. This heat provokes evaporation so that the tank pressure continuously increases. The tank used for this work represents a two layered glass dewar vessel equipped with a polyacetal tank lid carrying the instrumentation. Notwithstanding the fact that the dewar tank consists of a vacuum layer and is covered with a silvering layer on the inside, it is assumed that a certain amount of heat conducts from the lid through the glass tank walls to the liquid surface to cause evaporation. The total amount of heat that flows into the tank is measured to be approximately $\dot{Q}_{\text{heat}} \approx 6$ W for an open system. The thermodynamic consideration of the pressurized phase could shows that in fact $\dot{Q} \approx 10$ W for a closed system. In the full size application, the heat input strongly depends on the respective tank part (such as lower bulkhead, common bulkhead or upper bulkhead) and its insulation. Considering the ESC-B hydrogen tank compartment as shown in figure 5.1, the tank bottom as part of the common bulkhead design represents also the ceiling of the oxygen tank below. Therefore, the heat exchange here differs significantly from the heat exchange at the side walls for instance. However, the time scale for self-pressurization is directly linked to these heat fluxes and therefore on the efficiency of the applied insulation concept.

By considering the GN_2 pressurization experiments, the tank pressure is ramped up by injecting gaseous nitrogen from a high-pressure supply bottle. The time scale for external GN_2 pressurization is approximately $1/10$ of the time scale considered for self-pressurization. The current experimental results show that a certain amount of vapor re-condenses during the pressurization phase where warmer GN_2 is injected through the inlet. Studying the temperature distribution in the liquid for the corresponding initial pressure p_0^*, it turned out that the pressurization time scale has a major influence on the thermal stratification in the liquid. Confirming the theory by DAS & HOPFINGER [26], the highest temperature gradient in the liquid is found in the vicinity of the free liquid surface and can be expressed by the thermal boundary layer depth δ_t given by the theoretical expression provided in equation (2.91) by analogy of momentum diffusion [26]. Measurements in this work only represent a vague approximation of δ_t since the temperature sensor configuration in the liquid of the actual setup is only suitable to a limited extent resolving this length scale in the order of a few millimeters.

The thermal boundary layer depth δ_t has a major impact on the pressure drop showing the highest temperature gradient in the liquid. This length represents the transition where the colder liquid from the bulk mixes with the warmer liquid from the surface during sloshing. The saturation temperature at the surface is coupled to the tank pressure by the CLAUSIUS CLAPEYRON relation. Thus, the sudden decrease of the temperature and therefore of the tank pressure provoke condensation in the lower vapor strata. The higher the temperature gradient in δ_t, the higher the impact of the pressure drop. This can be measured by the pressure drop magnitude Δp_{mag}^* representing the maximum pressure loss during sloshing and by the pressure gradient $|\partial p^* / \partial t^*|$ representing the time scale on which the characteristic pressure drop occurs. With respect to the full size application where the tank consists only of an insulated aluminum shell, the heat fluxes entering the tank are assumed to be much higher causing more convective flow in the liquid. The GRASHOF number expressing thermal convection is assumed to be several decades higher in upper stage tanks than in the experiments conducted for this work as shown in figure 2.5 (A). This might be in contrast to the thermal stratification observed in the vacuum insulated dewar tank used for this work. Convection might disturb the development of the thermal boundary layer for some extent so that the expected pressure drop is of smaller impact.

Nevertheless, a stratified ullage is observed, while the pressure drop occurs. But the thermal stratification in the ullage particularly in the upper regions has only a minor influence on the heat transfer in the vicinity of the free surface. This is confirmed by regarding the energy balance during sloshing of the GN_2 pressurized tank indicating that the cooling-down of the ullage gas represents only 10% of the total pressure drop, while the remaining 90% is caused by condensation.

The application of helium as non-condensable pressurant gas leads to the expected experimental results. The basic conclusion here: The higher the helium concentration in the ullage, the smaller the pressure drop caused by condensation. For a helium/nitrogen system, a threshold helium ullage concentration of approximately $\chi_{GHe} \approx 0.73$ is measured where the pressure drop can still be observed. For exceeding $\chi_{GHe} \approx 0.73$, the tank pressure does not drop under the impact of sloshing. As shown in the energy balance for the GHe experiments, the condensation effects during sloshing are already stopped for a concentration of approximately $\chi_{GHe,\,f+p} \approx 0.6$. For increasing helium concentration, the pressure drop is driven by the ullage cooling-down only. Assuming sloshing in the lateral regime, it is expected that the threshold concentration can also be applied to suppress the pressure drop in the hydrogen tank of the full size application. Helium reduces the vapor partial pressure and therefore affects the phase change (condensation) of the propellant vapor phase. These thermodynamic properties are characterized by the JACOB number, which appears in a similar order of magnitude (difference of factor 2-3) for cryogenic systems including LN_2, LH_2, and LOX. Due to the application of helium, the propellant partial pressure decreases and therefore the saturation temperature as well. Thus, the characteristic temperature difference in the liquid Θ_L as parameter of the JACOB number decreases for approximately constant bulk temperatures. Assuming that the bulk temperature is similar to ϑ_{ref}, the characteristic temperature difference Θ_L is similar to the degree of subcooling in the liquid. The degree of subcooling is shown in figure B.4 (B) in the appendix where ϖ_{sub} is plotted for increasing helium concentrations.

The development of the actual pressure data as function of the required initial pressure allows the linear extrapolation for both, the pressure drop magnitude Δp^*_{mag} and the pressure gradient $|\partial p^*/\partial t^*|$. The results for the pressure drop magnitude are in good agreement to the data provided by previous experimental studies taken from the literature considering higher initial pressures and different fluids [39, 48, 26]. The thermal stratification within the liquid is expressed by the thermal boundary layer depth δ_t. This parameter is primary influenced by the pressurization history with respect to self-pressurization or external gas pressurization. By initiating the sloshing motion, this mainly impacts the mixing of the liquid at the free surface with colder liquid from the bulk resulting into the decrease of the saturation temperature and therefore to the decrease of the tank pressure. The reduction of the surface temperature provokes condensation in the vapor phase. The time scale how fast the tank pressure drops is directly linked to the excitation and therefore to the corresponding liquid motion provoking the thermal mixing. However, the actual data is compared to the theoretical approach given by DAS & HOPFINGER [26] to express the thermal mixing caused by sloshing by the effective diffusion coefficient D_e. The effective diffusion coefficient corresponds to the thermal diffusivity D_T in case of a motionless interface (static position). Considering the actual sloshing motion, the dimensionless diffusion coefficient for the GN_2 pressurization experiments gives $D_e'^* \approx 0.4 \ldots 1.4 \times 10^{-6}$. Since the thermal boundary layer depth also depends on the pressurization time scale, $D_e'^*$ linearly increases for an increasing initial pressure. This conclusion must be considered carefully in particular for splashing waves as they occur near resonance. Here, the splashing liquid from the free liquid surface may essentially influence the result and impede reliable predictions.

According to the presented results, the pressure drop magnitude $\Delta p^*_{\mathrm{mag}}$ does not depend on the excitation. Sloshing provokes thermal mixing independently from the wave amplitude. Merely the duration to reach the pressure minimum is impacted by the excitation showing higher pressure gradients for higher excitations. Confirmed by previous data [48, 39, 26], this assumption is applicable for the lateral sloshing regime but for splashing waves near resonance more investigations are necessary for reliable conclusions.

Actual and previous experimental results including liquid nitrogen (LN_2), liquid oxygen (LOX) and liquid hydrogen (LH_2) show that the behavior of cryogenic one-species systems can be predicted for a harmonic lateral excitation. On upper stages during ascent phase, the excitation is most typically activated by a sudden change of the lateral acceleration. Thus, the damping characteristics of the liquid and the tank design (common bulkhead dome, baffles) might influence the sloshing motion in the tank by a higher extent and therefore the pressure drop as well. The thermal stratification expressed by δ_t is also influenced by the heat fluxes entering the tank to provoke convective motion that might act as a thermal stirrer. Thus, the results in this work enhance the understanding of the pressure drop effect in order to provide guidelines to prevent this phenomenon for future cryogenic upper stage tanks.

Chapter 6

OUTLOOK

This chapter is dedicated to the future activities in the field of cryogenic propellant management particularly considering the free surface motion and other effects connected to this topic.

As described in the previous chapter, the highest uncertainty in the prediction of the viscous damping in upper stage tanks can be found in differences between the tank geometries in laboratory size experiments that mostly consist of simple cylindrical shapes and the full size application considering complex forms. It may be suggested to perform laboratory size cryogenic damping experiments in order to determine the damping parameters for the planned ESC-B upper stage tank setup as shown in figure 5.1 (A). The objective of this study might include the determination of the geometry related dome factor for the characteristic convex bottom shape configuration given by the common bulkhead. These tests could be carried out in a 1/10 scaled model of the full size application made of acrylic glass using storable liquids under ambient pressure conditions.

For security reason, the actual experimental setup introduced in this work allows only initial pressures up to a maximum value of $p_{\mathrm{max}} = 160$ kPa. Previous and actual cryogenic upper stage tank pressures range up to $p = 450$ kPa. To confirm the accuracy of the linear extrapolation presented in figures 4.37 and 4.35, it might be adequate to continue the current experiments with a similar setup appropriate to be used for higher initial pressures. Additionally, it might be of interest to study the usage of ring baffles on the impact of the pressure drop effect. Ring baffles are preferably applied in rocket tanks to suppress liquid sloshing.

MORAN et al. [48] provided experimental data based on studies in a $R = 0.75$ m spherical tank that was filled with liquid hydrogen to various fill levels. It might be of interest to repeat their experiments in a cylindrical tank such as it was adopted here using hydrogen and a similar excitation. As pointed out in this work, the tank shape is of particular importance to significantly influence the sloshing characteristics in the tank and therefore the heat and mass transfer at the liquid/vapor interface as well. This might provide further information about the discrepancy of the pressure gradient results determined by MORAN et al. [48].

To investigate further mission phases might include the consideration of experiments under micro-gravity conditions as well. Particularly for the engine cut-off and restart, the conditions in the tank are not fully understood. In terms of liquid sloshing, the variation of the acceleration in direction of the rockets rotational axis might end up in a different sloshing mode that is related to the first axial sloshing mode ($m = 1$, $n = 1$) as illustrated in figure 2.10. As previously observed by WARD *et al.* [57] in the S-IVB stage hydrogen tank of the United States Saturn-1B space launcher, the liquid performs large axial movement under the lack of gravitational acceleration in orbit, while executing a stopping maneuver. The increasingly larger liquid surface may result in an increase of the pressure drop. This high interesting area also represent a field of further investigation.

Chapter 7

SUMMARY

This work is dedicated to cryogenic liquid sloshing in accordance to the periodic propellant motion as it is expected on upper stages particularly during the ascent phase after lift-off. The objectives of this work include studies to investigate the damping behavior of a cryogenic liquid as well as to enhance the understanding of the characteristic pressure drop phenomenon that occurs under the impact of propellant sloshing to allow predictions for future cryogenic upper stages.

In the frame of this work, laterally excited sloshing tests are performed in a $R = 0.145$ m cylindrical dewar tank with a spherical bottom geometry using liquid nitrogen (LN$_2$) as substitute for the cryogenic upper stage fuel liquid hydrogen (LH$_2$). The excitation corresponding to a quasi harmonic oscillation is defined by a frequency of $f = 1.4$ Hz, while the amplitude is set to $y_A = 0.01$ m. This is appropriate to excite the liquid in the first asymmetric sloshing mode that is also assumed on the upper stage during the ascent phase.

Decay experiments are performed in order to determine the damping characteristic of the actual cryogenic system. The results are compared to the theory developed by MIKISHEV et $al.$ [45] and STEPHENS et $al.$ [55] in order to ensure its validity also for the cryogenic (very low viscosity) domain. In terms of the damping ratio γ, STEPHENS et $al.$could reduce the damping characteristics to a relation that includes the GALILEI number satisfying $\gamma = K \, \mathbf{Ga}^{-1/4}$ for higher fill level $H/R \geq 1$. For a cylindrical geometry, the damping coefficient gives $K = 0.83$ [27, 37]. The cryogenic results for various fill level presented in this work show good agreement to this prediction based on the potential theory modified by a vector potential function to consider the viscous properties of the liquid [37].

Furthermore, it is assumed that sloshing propellant may influence the interactions at the liquid/vapor interface at the free surface causing condensation. In order to reproduce the characteristic pressure drop as a result of the fluid motion, cryogenic sloshing experiments are performed with LN$_2$ under variation of the initial condition by applying three different methods to pressurize the system. This includes self-pressurization, external nitrogen (GN$_2$) pressurization and external helium (GHe) pressurization.

After pressurization, the liquid appears in a certain state where the free surface has saturation temperate and is therefore in equilibrium to the vapor phase above. The liquid bulk is subcooled with respect to the temperature at the surface. The thermal boundary layer between the free surface and the liquid bulk is defined by the highest temperature gradient in the liquid. The ullage temperature is linearly stratified. This stratification is limited between the saturation temperature at the free surface and the temperature at the tank lid.

However, under the impact of sloshing, the liquid of the thermal boundary layer in the vicinity of the free surface gets mixed with colder liquid from the bulk effecting the decrease of the temperature at the free surface. Based on the CLAUSIUS CLAPEYRON law, this provokes the reduction of the saturation temperature and therefore the tank pressure to decreases as well. Thus, the sudden decrease of the tank pressure leads to vapor condensation particularly in the vicinity of the free surface. Experimental results confirm that the impact of the pressure drop depends significantly on the initial pressure as well as on the pressurization duration, which is a factor 10 higher in case of self-pressurization as for GN_2/GHe pressurization. By varying the helium concentration in the ullage, the application of a non-condensable inert gas (GHe) decreases the impact of the pressure drop. For a helium concentration $\chi_{He} \geq 0.73$ the pressure drop could be entirely stopped, so that the tank pressure further increases even after starting the excitation.

The pressure drop phenomenon is characterized by two according numbers; the pressure drop magnitude defined as difference between the initial pressure and the minimum pressure after sloshing as well as the pressure gradient defined as pressure loss per time expressing the temporal behavior of the excited cryogenic liquid. Predictions for the full size application are based on linear extrapolations of the experimental data (pressure drop magnitude and pressure gradient) obtained from this work. The predictions are confirmed by data taken from the literature [39, 48, 26] performed under higher pressure conditions for a large amplitude sloshing regime near the first natural frequency. The model of DAS and HOPFINGER [26] is applied to correlate the high pressure results [39, 48, 26] also for the smaller excitation as considered here. Independent from the sloshing regime, the magnitude of the pressure drop can be predicted by knowing the initial pressure and the ullage volume. The pressure gradient essentially depends on the sloshing regime showing significantly higher values for higher wave amplitudes. Particularly in the vicinity of the first natural frequency, this provokes stronger interactions at the phase boundary.

Bibliography

[1] H. N. Abramson. The dynamic behavior of liquids in moving containers. Technical Report NASA-SP-106, Southwest Research Institute, San Antonio, TX, USA, 1967.

[2] H. N. Abramson, W. H. Chu, and D. D. Kana. Some studies of nonlinear lateral sloshing in rigid containers. *J. Applied Mech.*, pages 777–784, 1966.

[3] T. Arndt. Sloshing tests considering lateral excited liquid in cylindrical propellant vessels. Technical report, ZARM, University of Bremen, Bremen, Germany, January 2007.

[4] T. Arndt. Stratification and destratification of liquid nitrogen LN2 in closed cylindrical propellant tanks. Technical report, ZARM, University of Bremen, Bremen, Germany, September 2009.

[5] T. Arndt and M. Dreyer. Damping behavior of sloshing liquid in laterally excited cylindrical propellant vessels. In *43rd AIAA/ASME/SAE/ASEE Joint Propulsion Conference and Exhibit*, number AIAA 2007-5162, Cincinnati, OH, USA, July 2007.

[6] T. Arndt and M. Dreyer. Damping behavior of sloshing liquid in laterally excited cylindrical propellant vessels. *J. Spacecraft Rockets*, 45(5):1085–1088, 2008.

[7] T. Arndt, M. Dreyer, P. Behruzi, and M. Winter. Laterally excited sloshing tests with liquid nitrogen. In *44th AIAA/ASME/SAE/ASEE Joint Propulsion Conference and Exhibit*, number AIAA 2008-4551, Hartford, CT, USA, July 2008.

[8] T. Arndt, M. Dreyer, P. Behruzi, M. Winter, and A. van Foreest. Cryogenic sloshing tests in a pressurized cylindrical reservoir. In *45th AIAA/ASME/SAE/ASEE Joint Propulsion Conference and Exhibit*, number AIAA 2009-4860, pages 1–10, Denver, CO, USA, August 2009.

[9] J. C. Aydelott. Effect of gravity on self-pressurization of spherical liquid-hydrogen tankage. Technical Report NASA-TN-D-4171, NASA Glenn Research Center, Cleveland, OH, USA, December 1967.

[10] J. C. Aydelott. Normal gravity self-pressurization of 9-inch- 23 cm diameter spherical liquid hydrogen tankage. Technical Report NASA-TN-D-4171, NASA Glenn Research Center, Cleveland, OH, USA, October 1967.

[11] H. D. Baehr and S. Kabelac. *Thermodynamik*, volume 12. Springer Verlag, 2006.

[12] H. D. Baehr and K. Stephan. *Wärme- und Stoffübertragung*, volume 5. Springer Verlag, 2006.

[13] T. Bailey, R. VanderKoppel, G. Skartvedt, and T. Jefferson. Cryogenic propellant stratification analyisis and test data correlation. *AIAA Journal*, 1(7):1657–1659, July 1963.

[14] D. O. Barnett. Liquid nitrogen stratification analysis and experiments in a partially filled, spherical container. Technical report, Northrop Space Laboratories, Huntsville, AL, USA, 1967.

[15] H. F. Bauer. Fluid oscillations in the containers of a space vehicle and their influence on stability. Technical Report TR R-187, NASA, George C. Marshall Space Flight Center, Huntsville, AL, USA, February 1964.

[16] H. F. Bauer. Hydroelastische Schwingungen im aufrechten Kreiszylinder. *Z. Flugwiss. 18*, 4:117 – 134, 1970.

[17] K. P. Behruzi and G. Netter. PMD design for upper stages. *4th International Conference on Launcher Technology: Space Launcher Liquid Propulsion, Liege, Belgium*, December 2002.

[18] U. L. Berkes. The Ariane 5 ECA Heavy-Lift Launcher. Ariane Project Department, Directorate of Launchers, ESA, Paris, France, August 2005.

[19] R. B. Bird, W. E. Stewart, and E. N. Lightfood. *Transport Phenomena*. John Wiley, New York, 1960.

[20] M. Broda. Experimentelle und numerische Untersuchungen zum dreidimensionalen Flüssigkeitsschwappen in einem zylindrischen Behälter bei lateraler Anregung und grossen Bond-Zahlen. Master's thesis, University of Bremen, Bremen, Germany, January 2006.

[21] I. N. Bronstein and K. A. Semendjajew. *Handbook of Mathematics*. Springer, 4 edition, Jun. 2004.

[22] K. M. Case and W. C. Parkinson. Damping of surface waves in an incompressible liquid. *J. Fluid Mech.*, 2:172–184, 1957.

[23] S. Chandrasekhar. *Hydrodynamic and Hydromagnetic Stability*. Dover Publications Inc., New York, 1961.

[24] J. A. Clark. A review of pressurization, stratification and interfacial phenomena. *Adv. Cryo. Eng.*, 10:259 – 283, 1965.

[25] S. P. Das and E. J. Hopfinger. Parametrically forced gravity waves in a circular cylinder and finite-time singularity. *J. Fluid Mech.*, 599:205–228, 2008.

[26] S. P. Das and E. J. Hopfinger. Mass transfer enhancement by gravity waves at a liquid vapour interface. *Int. J. Heat Mass Tran.*, 52(5-6):1400–1411, February 2009.

[27] F. T. Dodge. *The new dynamic behavior of liquids in moving containers*. Southwest Research Institute, San Antonio, TX, USA, 2000.

[28] N. T. Van Dresar. PVT gauging with liquid nitrogen. *Cryogenics*, 46:118 – 125, 2006.

[29] M. E. Dreyer. *Free Surface Flows under Compensated Gravity Conditions*, volume 221 of *Springer Tracts in Modern Physics*. Springer Verlag, 2007.

[30] M. E. Dreyer, U. Schmid, and M. Stange. Test parameters for propellant settling time determination EPSV. Technical report, Center of Applied Space Technology and Microgravity (ZARM), University of Bremen, Bremen, Germany, January 2000.

[31] G. Grayson, A. Lopez, and F. Chandler. Cryogenic tank modeling for the saturn AS-203 experiment. In *42nd AIAA/ASME/SAE/ASEE Joint Propulsion Conference and Exhibit*, number AIAA-2006-5258, Sacramento, CA, USA, July 2006.

[32] G. Grayson, A. Lopez, F. Chandler, L. Hastings, A. Hedayat, and J. Brethour. CFD modeling of helium pressurant effects on cryogenic tank pressure rise rates in normal gravity. In *43rd AIAA/ASME/SAE/ASEE Joint Propulsion Conference and Exhibit*, number AIAA 2007-5524, Cincinnati, OH, USA, July 2007.

[33] D. Haake. Dreidimensionales Flüssigkeitsschwappen in einem zylindrischen Behälter bei lateraler Anregung und großen Bondzahlen. Technical report, University of Bremen, Bremen, Germany, August 2002.

[34] T. Himeno, Y. Umemura, D. Sugimori, S. Uzawa, and T. Watanabe. Investigation on heat exchange enhanced by sloshing. In *45th AIAA/ASME/SAE/ASEE Joint Propulsion Conference and Exhibit*, number AIAA 2009-5397, Denver, CO, USA, August 2009.

[35] T. Himeno, Y. Umemura, T. Watanabe, and H. Aoki. Preliminary investigation on heat exchange enhanced by sloshing. In *44th AIAA/ASME/SAE/ASEE Joint Propulsion Conference and Exhibit*, number AIAA 2008-4550, Hartford, CT, USA, July 2008.

[36] E. J. Hopfinger and S. P. Das. Mass transfer enhancement by capillary waves at a liquid-vapor interface. *Exp. Fluids*, 46(4):597–605, 2009.

[37] R. A. Ibrahim. *Liquid sloshing dynamics*. Cambridge University Press, 2005.

[38] S. P. Kumar, B. V. S. S. S. Prasad, G. Venkatarathnam, K. Ramamurthi, and S. S. Murthy. Influence of surface evaporation on stratification in liquid hydrogen tanks of different aspect ratios. *Int. J. Hydrogen Energ.*, 32(12):1954–1960, August 2007.

[39] J. Lacapere, B. Vieille, and B. Legrand. Experimental and numerical results of sloshing with cryogenic fluids. In *2nd European Conference for Aerospace Sciences (EUCASS)*, pages 1–8, 2007.

[40] H. Lamb. *Hydrodynamics*. Cambridge University Press, Cambridge, 7 edition, 1932.

[41] E. Lemmon, A. Peskin, M. McLinden, and D. Friend. NIST12 thermodynamic and transport properties of pure fluids (standard reference database 12, version 5.0). Software, 2000.

[42] T. N. Lovrich and S. H. Schwartz. Development of thermal stratification and destratification scaling concepts. Technical Report NASA-CR-143944, NASA, October 1975.

[43] K. Magnus. *Schwingungen*. Teubner Studienbücher Mechanik, Stuttgart, 1986.

[44] E. Messerschmid and S. Fasoulas. *Raumfahrtsysteme*. Springer-Verlag, Berlin Heidelberg, 2009.

[45] G. N. Mikishev and N. Y. Dorozhkin. An experimental investigation of free oscillations of a liquid in containers. *Izv. Akad. Nauk. SSSR Otd. Tekhn. Nauk. Mekh. I Mashinostr.*, 4:48–83, 1961. (Translated into English by D. D. Kana, SwRI, June 30, 1963).

[46] J. W. Miles. On the sloshing of liquid in a cylindrical tank. Technical Report GM-TR-18, Guided Missile Research Division, USA, April 1956.

[47] J. W. Miles. Resonantly forced surface waves in a circular cylinder. *J. Fluid Mech.*, 149:15–31, 1984.

[48] M. E. Moran, N. B. McNelis, M. T. Kudlac, M. S. Haberbusch, and G. A. Santornino. Experimental results of hydrogen slosh in a 62 cubic foot (1750 liter) tank. In *30th AIAA/ASME/SAE/ASEE Joint Propulsion Conference*, number AIAA-94-3259, pages 1–10, Indianapolis, IN, USA, June 1994.

[49] H. R. Rath. Mechanik IV. University of Bremen, Bremen, Germany, 1992.

[50] E. Ring. *Rocket Propellant and Pressurization Systems*. Prentice-Hall, Inc., Englewood Cliffs, NJ, USA, 1964.

[51] A. Royon-Lebeaud. *Ballottement des liquides dans les réservoirs cylindriques soumis à une oscillation harmonique*. PhD thesis, Université Joseph Fourier, Laboratoire des Ecoulements Géophysiques et Industriels, France, March 2005.

[52] A. Royon-Lebeaud, E. J. Hopfinger, and A. Cartellier. Liquid sloshing and wave breaking in circular and square-base cylindrical containers. *J. Fluid Mech.*, 577:467–494, 2007.

[53] R. H. Scanlan and R. Rosenbaum. *Introduction to the Study of Aircraft Vibration and Flutter*. The Macmillan Company, USA, 1951.

[54] P. Stephan, K. Schaber, K. Stephan, and F. Mayinger. *Thermodynamik*, volume 18. Springer Verlag, 2008.

[55] D. G. Stephens, H. W. Leonard, and T. W. Perry Jr. Investigations of the damping of liquids in right-circular cylindrical tanks, including the effects of time-variant liquid depth. Technical report, NASA, Langley Research Center, Langley Station, Hampton, VA, USA, 1962.

[56] J. Verne. *Von der Erde zum Mond*. Verlag Bärmeier & Nickel, 1865.

[57] W. D. Ward, L. E. Tool, C. A. Ponder, M. E. Meadows, C. W. Simmons, J. H. Lytle, J. M. McDonald, and M. Kavanaugh. Evaluation of AS-203 low gravity orbital experiment. Technical Report HSM-R421-67, Contract NAS8-4016, Chrysler Corporation, Space Division, USA, January 1967.

Appendix A

Tables

Table A.1: Properties of duran glass from which material the dewar is made of. The tank density is approximately independent from the temperature and gives $\varrho_{\text{dewar}} = 2230 \text{ kg m}^{-3}$.

ϑ [K]	k [W m^{-1} K^{-1}]	c_v [J kg^{-1} K^{-1}]
70	0.454590356	200.9
100	0.622798260	299.6
130	0.746529247	392.9
160	0.844451971	480.8
190	0.925496552	563.3
220	0.994634751	640.4
250	1.054920969	712.1
280	1.108366777	778.4
310	1.156367495	839.3

Table A.2: Durations for self-pressurization and GN$_2$ pressurization experiments.

p_0 [kPa]		120	130	140	150	160
t_p [s]	selfpressurization	740	1180	1800	2000	2432
	GN$_2$ pressurization	72	126	180	228	294

Table A.3: Quantity of GN_2 brought in the system while pressurizing the test tank for a mass flow rate of $\dot{m}_{GN2} = 0.105$ g s^{-1}.

p_0 [kPa]	t_p [s]	m_{GN2} [g]
120	72	0.00756
130	126	0.01323
140	180	0.0189
150	228	0.02394
160	294	0.03087

Table A.4: Quantity of GHe brought in the system while flushing and pressurizing the test tank for a mass flow rate of $\dot{m}_{GHe} = 0.021$ g s^{-1}.

flushing		pressurization		total	
t_f [s]	m_{GHe} [g]	t_p [s]	m_{GHe} [g]	t_{f+p} [s]	m_{GHe} [g]
18	0.38	127	2.69	145	3.07
48	1.02	124	2.63	172	3.65
78	1.65	120	2.55	198	4.20
108	2.29	118	2.50	226	4.79
138	2.93	118	2.50	256	5.43
168	3.56	109	2.31	227	5.87
198	4.20	101	2.14	299	6.34
228	4.84	99	2.10	327	6.94

Table A.5: Quantity to determine the molar fractions of gaseous helium and nitrogen after flushing.

t_f	m_GHe	m_GN2	V_GHe	V_GN2	n_GHe	n_GN2	χ_GHe	χ_GN2	$\bar{M}$
[s]	[g]	[g]	[m³]	[m³]	[mol]	[mol]	[mol mol⁻¹]	[mol mol⁻¹]	[kg mol⁻¹]
18	0.38	38.47	0.0015	0.0223	0.1	1.37	0.07	0.93	26.33
48	1.02	34.33	0.0039	0.0199	0.25	1.22	0.17	0.83	23.93
78	1.65	30.19	0.0063	0.0175	0.41	1.08	0.28	0.72	21.29
108	2.29	25.88	0.0088	0.015	0.57	0.92	0.38	0.62	18.89
138	2.93	21.56	0.0113	0.0125	0.73	0.77	0.49	0.51	16.25
168	3.56	17.42	0.0137	0.0101	0.89	0.62	0.59	0.41	13.85
198	4.2	13.11	0.0162	0.0076	1.05	0.47	0.69	0.31	11.45
228	4.84	8.97	0.0186	0.0052	1.21	0.32	0.79	0.21	9.04

Table A.6: Quantity to determine the molar fractions of gaseous helium and nitrogen after flushing and pressurization.

t_{f+p}	m_GHe	m_GN2	V_GHe	V_GN2	n_GHe	n_GN2	χ_GHe	χ_GN2	$\bar{M}$
[s]	[g]	[g]	[m³]	[m³]	[mol]	[mol]	[mol mol⁻¹]	[mol mol⁻¹]	[kg mol⁻¹]
145	3.07	29.71	0.0091	0.0147	0.66	1.06	0.38	0.62	18.83
172	3.65	26.22	0.0108	0.0129	0.78	0.94	0.46	0.54	17.09
198	4.2	22.91	0.0125	0.0113	0.9	0.82	0.52	0.48	15.44
226	4.79	19.36	0.0142	0.0096	1.03	0.69	0.6	0.4	13.67
256	5.43	15.51	0.0161	0.0077	1.16	0.55	0.68	0.32	11.75
277	5.87	12.86	0.0174	0.0063	1.26	0.46	0.73	0.27	10.42
299	6.34	10.03	0.0188	0.0049	1.36	0.36	0.79	0.21	9.01
327	6.94	6.41	0.0206	0.0032	1.49	0.23	0.87	0.13	7.21

Table A.7: Tank condition after flushing and pressurizing the test tank.

t_f [s]	ϑ_{sat} [K]	p_{GN2} [kPa]	p_{GHe} [kPa]
18	79.69	132.28	7.82
48	79.36	127.52	12.71
78	79.00	122.47	17.59
108	78.64	117.58	22.56
138	78.30	113.04	27.04
168	77.95	108.57	31.87
198	77.68	104.92	35.13
228	77.60	104.26	36.14

Table A.8: Reference quantities for data scaling (self-pressurization).

p_0 [kPa]	ϑ_{ref} [K]	$\vartheta_{\mathrm{sat},0}$ [K]	ϑ_{lid} [K]	$\varrho_{U,\mathrm{ref}}$ [kg m^{-3}]	R_s [J kg^{-1} K^{-1}]
120	77.35	78.82	285	4.61	296.8
130	77.35	79.53	288	4.61	296.8
140	77.35	80.21	286	4.61	296.8
150	77.35	80.84	285	4.61	296.8
160	77.35	81.45	284	4.61	296.8

Table A.9: Reference quantities for data scaling (GN_2 pressurization).

p_0 [kPa]	ϑ_{ref} [K]	$\vartheta_{\text{sat},0}$ [K]	ϑ_{lid} [K]	$\varrho_{U,\text{ref}}$ [kg m^{-3}]	R_s [J kg^{-1} K^{-1}]
120	77.35	78.82	290	4.61	296.8
130	77.35	79.53	288	4.61	296.8
140	77.35	80.21	285	4.61	296.8
150	77.35	80.84	287	4.61	296.8
160	77.35	81.45	286	4.61	296.8

Table A.10: Reference quantities for data scaling (GHe pressurization).

p_0 [kPa]	ϑ_{ref} [K]	$\vartheta_{\text{sat},0}$ [K]	ϑ_{lid} [K]	$\varrho_{U,\text{ref}}$ [kg m^{-3}]	R_s [J kg^{-1} K^{-1}]
18	77.35	80.21	283	4.61	296.8
48	77.35	80.21	284	4.61	296.8
78	77.35	80.21	285	4.61	296.8
108	77.35	80.21	283	4.61	296.8
138	77.35	80.21	282	4.61	296.8
168	77.35	80.21	281	4.61	296.8
198	77.35	80.21	283	4.61	296.8
228	77.35	80.21	282	4.61	296.8

Table A.11: Tabular data of the liquid temperature sensors during self-pressurization experiments. Please note, that the data provided for $z_{\mathrm{pos}} = 0.29$ m (at the free surface) is not measured but determined based on equation (2.82).

initial pressure level	$p_0 = 120$ kPa	130	140	150	160
start press	$p = 101.12$ kPa	101.23	102.55	102.64	102.32
z_{pos} [m]	ϑ_L [K]				
0.29	77.34	77.35	77.46	77.46	77.44
0.278	77.51	77.48	77.55	77.55	77.48
0.253	77.54	77.51	77.53	77.53	77.5
0.203	77.66	77.65	77.66	77.66	77.62
0.103	77.58	77.56	77.55	77.57	77.54
0.003	77.55	77.53	77.54	77.55	77.54
end press	$p = 120$ kPa	130	140	150	160
0.29	78.82	79.53	80.21	80.84	81.45
0.278	78.09	78.44	78.89	79.28	79.73
0.253	77.72	77.82	77.97	78.03	78.16
0.203	77.79	77.86	77.96	77.99	78.02
0.103	77.67	77.72	77.84	77.86	77.90
0.003	77.66	77.71	77.82	77.84	77.88
end slosh	$p = 114.88$ kPa	120.14	125.15	130.83	135.1
0.29	78.44	78.83	79.19	79.59	79.88
0.278	78.46	78.87	79.23	79.55	79.87
0.253	78.27	78.65	79.02	79.37	79.71
0.203	77.85	77.99	78.2	78.39	78.48
0.103	77.72	77.81	77.91	77.95	77.99
0.003	77.68	77.79	77.9	77.94	77.97

Table A.12: Tabular data of the liquid density during self-pressurization experiments taken from [41].

initial pressure level	$p_0 = 120$ kPa	130	140	150	160
start press	$p = 101.12$ kPa	101.23	102.55	102.64	102.32
z_pos [m]		ϱ_L [kg m^{-3}]			
0.29	806.16	806.12	805.62	805.59	805.71
0.278	805.39	805.54	805.2	805.22	805.52
0.253	805.23	805.36	805.29	805.3	805.44
0.203	804.7	804.76	804.7	804.71	804.9
0.103	805.08	805.16	805.19	805.12	804.79
0.003	805.22	805.29	805.24	805.26	805.26
end press	$p = 120$ kPa	130	140	150	160
0.29	799.39	796.1	792.98	790	787.15
0.278	802.78	801.19	799.11	797.32	795.26
0.253	804.49	804.05	803.39	803.1	802.53
0.203	804.14	803.85	803.4	803.31	803.18
0.103	804.7	804.49	803.97	803.9	803.76
0.003	804.73	804.54	804.07	803.98	803.82
end slosh	$p = 114.88$ kPa	120.14	125.15	130.83	135.1
0.29	801.15	799.35	797.68	795.84	794.49
0.278	801.06	799.17	797.5	796.03	794.54
0.253	801.94	800.17	798.49	796.85	795.31
0.203	803.84	803.2	802.29	801.4	801
0.103	804.44	804.05	803.61	803.43	803.25
0.003	804.65	804.12	803.65	803.49	803.35

Table A.13: Tabular data of the liquid enthalpy during self-pressurization experiments taken from [41].

initial pressure level	$p_0 = 120$ kPa	130	140	150	160
start press	$p = 101.12$ kPa	101.23	102.55	102.64	102.32
z_{pos} [m]	$h_{e,L}$ [J kg^{-1}]				
0.29	-122053.5	-122035.4	-121809.6	-121794.4	-121848.2
0.278	-121704.5	-121772.1	-121618.1	-121628.6	-121761.8
0.253	-121630.6	-121692.2	-121659.2	-121663.4	-121729
0.203	-121394.7	-121421.4	-121394.5	-121398.9	-121485
0.103	-121562.9	-121599.9	-121616.1	-121582.7	-121431.6
0.003	-121628.6	-121659.3	-121636.8	-121644	-121644
end press	$p = 120$ kPa	130	140	150	160
0.29	-119010.6	-117539.8	-116149.3	-114829.3	-113777.5
0.278	-120509.6	-119786.7	-118846.1	-118039.7	-117314.8
0.253	-121270	-121058.8	-120745.3	-120597.3	-120524
0.203	-121116.8	-120966.8	-120749.4	-120691.4	-120812.1
0.103	-121366	-121252.9	-121004.8	-120955	-121069.5
0.003	-121378.3	-121275.3	-121045.7	-120989.8	-121096.1
end slosh	$p = 114.88$ kPa	120.14	125.15	130.83	135.1
0.29	-119798.9	-118989.4	-118242.4	-117421.5	-116821.3
0.278	-119758	-118908.3	-118164.8	-117504.7	-116844.6
0.253	-120147	-119353.3	-118602	-117868	-117183.6
0.203	-120991.5	-120699.7	-120283.1	-119876.2	-119691.1
0.103	-121261.3	-121077.8	-120872	-120780.2	-120693.5
0.003	-121353.2	-121108.5	-120886.3	-120808.8	-120736.4

Table A.14: Tabular data of the liquid mass during self-pressurization experiments.

initial pressure level		$p_0 = 120$ kPa	130	140	150	160
start press		$p =101.12$ kPa	101.23	102.55	102.64	102.32
$V_{L,k}$ [m^3]	z_{pos} [m]		$m_{L,k}$ [kg]			
0.0004	0.29	0.3195	0.3195	0.3193	0.3193	0.3193
0.00122	0.278	0.9842	0.9843	0.9839	0.9839	0.9843
0.00248	0.253	1.9945	1.9948	1.9947	1.9947	1.995
0.00762	0.203	6.1302	6.1307	6.1302	6.1303	6.1317
0.0026	0.103	2.0916	2.0919	2.092	2.0918	2.0909
0.00112	0.003	0.9048	0.9049	0.9048	0.9049	0.9049
end press		$p = 120$ kPa	130	140	150	160
0.0004	0.29	0.3168	0.3155	0.3143	0.3131	0.312
0.00122	0.278	0.981	0.979	0.9765	0.9743	0.9718
0.00248	0.253	1.9927	1.9916	1.99	1.9892	1.9878
0.00762	0.203	6.1259	6.1237	6.1203	6.1196	6.1186
0.0026	0.103	2.0907	2.0901	2.0888	2.0886	2.0882
0.00112	0.003	0.9043	0.904	0.9035	0.9034	0.9032
end slosh		$p = 114.88$ kPa	120.14	125.15	130.83	135.1
0.0004	0.29	0.3175	0.3168	0.3161	0.3154	0.3149
0.00122	0.278	0.9789	0.9766	0.9745	0.9727	0.9709
0.00248	0.253	1.9864	1.982	1.9778	1.9738	1.97
0.00762	0.203	6.1236	6.1188	6.1118	6.105	6.102
0.0026	0.103	2.09	2.089	2.0878	2.0874	2.0869
0.00112	0.003	0.9042	0.9036	0.903	0.9029	0.9027

Table A.15: Tabular data of the ullage temperature sensors during self-pressurization experiments. Please note, that the data provided for $z_{\mathrm{pos}} = 0.29$ m (at the free surface) is not measured but determined based on equation (2.82).

initial pressure level	$p_0 = 120$ kPa	130	140	150	160
start press	$p = 101.12$ kPa	101.23	102.55	102.64	102.32
z_{pos} [m]	ϑ_U [K]				
0.635	285	288	286	285	284
0.484	186.62	201.3	199.11	198.5	187.4
0.434	157.06	167.87	164.9	171.02	162.42
0.384	129.07	134.97	130.88	141.88	139.32
0.334	107.37	108.99	105.46	115.34	117.22
0.29	77.34	77.35	77.46	77.46	77.44
end press	$p = 120$ kPa	130	140	150	160
0.635	285	288	286	285	284
0.484	190.76	208.66	208.36	208.53	208.54
0.434	162.1	175.81	175.38	178.94	178.36
0.384	135.07	144.97	144.84	150.61	151.23
0.334	113.48	119.47	119.58	124.4	126.36
0.29	78.82	79.53	80.21	80.84	81.45
end slosh	$p = 114.88$ kPa	120.14	125.15	130.83	135.1
0.635	285	288	286	285	284
0.484	207.5	210.28	210.32	210.97	208.73
0.434	173.93	177.28	177.33	180.73	177.82
0.384	141.12	144.52	144.72	149.11	146.52
0.334	103.88	106.5	105.71	108.17	106.78
0.29	78.44	78.83	79.19	79.59	79.88

Table A.16: Tabular data of the ullage density during self-pressurization experiments taken from [41].

initial pressure level	$p_0 = 120$ kPa	130	140	150	160
start press	$p = 101.12$ kPa	101.23	102.55	102.64	102.32
z_{pos} [m]	ϱ_U [kg m^{-3}]				
0.635	1.2	1.18	1.21	1.21	1.21
0.484	1.83	1.7	1.74	1.75	1.84
0.434	2.18	2.04	2.1	2.03	2.13
0.384	2.66	2.55	2.66	2.45	2.49
0.334	3.23	3.18	3.33	3.04	2.98
0.29	4.6	4.61	4.66	4.67	4.65
end press	$p = 120$ kPa	130	140	150	160
0.635	1.42	1.52	1.65	1.77	1.9
0.484	2.13	2.1	2.27	2.43	2.59
0.434	2.51	2.5	2.7	2.84	3.04
0.384	3.02	3.05	3.29	3.38	3.6
0.334	3.62	3.72	4.01	4.13	4.33
0.29	5.39	5.81	6.22	6.63	7.04
end slosh	$p = 114.88$ kPa	120.14	125.15	130.83	135.1
0.635	1.36	1.41	1.47	1.55	1.6
0.484	1.87	1.93	2.01	2.09	2.19
0.434	2.23	2.29	2.39	2.45	2.57
0.384	2.76	2.82	2.94	2.98	3.13
0.334	3.8	3.88	4.07	4.16	4.36
0.29	5.18	5.4	5.61	5.84	6.02

Table A.17: Tabular data of the ullage enthalpy during self-pressurization experiments taken from [41].

initial pressure level	$p_0 = 120$ kPa	130	140	150	160
start press	$p = 101.12$ kPa	101.23	102.55	102.64	102.32
z_pos [m]		$h_{e,U}$ $[\mathrm{J\,kg^{-1}}]$			
0.635	295573.9	298697.7	296611.8	295570.2	294529.6
0.484	193030.5	208353.8	206056.9	205425.1	193838.5
0.434	162114.5	173433.7	170317.4	176719.1	167723.3
0.384	132722.2	138937.5	134619.2	146181.9	143495.3
0.334	109753.1	111475.7	107700.9	118198.2	120199.2
0.29	77144.8	77151.5	77235.1	77240.7	77220.8
end press	$p = 120$ kPa	130	140	150	160
0.635	295527.4	298628.3	296520.2	295453.5	294386.4
0.484	197253.1	215902	215543.4	215670.5	215640.9
0.434	167257.6	181559	181041.7	184716.3	184043.2
0.384	138853.1	149194.7	148965.8	154959.8	155532.1
0.334	115990.8	122234.7	122233.7	127239.1	129216.4
0.29	78250.1	78767.0	79244.9	79688.7	80102.5
end slosh	$p =114.88$ kPa	120.14	125.15	130.83	135.1
0.635	295540	298652.1	296556.5	295500.7	294448.2
0.484	214754.5	217632.2	217656.9	218305	215952
0.434	179679.9	183152.5	183174.5	186698.8	183627.2
0.384	145277.8	148801.7	148973.7	153538	150777.2
0.334	105816.5	108541.1	107611.5	110169	108606.9
0.29	77968.4	78257.6	78521.5	78808.1	79015.3

Table A.18: Tabular data of the ullage mass during self-pressurization experiments.

initial pressure level		$p_0 = 120$ kPa	130	140	150	160
start press		$p = 101.12$ kPa	101.23	102.55	102.64	102.32
$V_{U,k}$ [m^3]	z_{pos} [m]			$m_{U,k}$ [kg]		
0.00832	0.635	0.00995	0.00986	0.01006	0.0101	0.01011
0.0033	0.484	0.00605	0.00561	0.00574	0.00577	0.00609
0.0033	0.434	0.0072	0.00674	0.00695	0.0067	0.00704
0.0033	0.384	0.0088	0.00841	0.0088	0.00811	0.00823
0.0031	0.334	0.01001	0.00987	0.01035	0.00943	0.00925
0.00145	0.29	0.00669	0.0067	0.00678	0.00678	0.00676
end press		$p = 120$ kPa	130	140	150	160
0.00832	0.635	0.01181	0.01266	0.01373	0.01477	0.01581
0.0033	0.484	0.00702	0.00695	0.0075	0.00803	0.00856
0.0033	0.434	0.00828	0.00826	0.00893	0.00937	0.01003
0.0033	0.384	0.00998	0.01006	0.01085	0.01118	0.01188
0.0031	0.334	0.01124	0.01156	0.01245	0.01281	0.01346
0.00145	0.29	0.00783	0.00844	0.00904	0.00963	0.01023
end slosh		$p = 114.88$ kPa	120.14	125.15	130.83	135.1
0.00832	0.635	0.01131	0.0117	0.01227	0.01288	0.01334
0.0033	0.484	0.00617	0.00637	0.00664	0.00692	0.00722
0.0033	0.434	0.00738	0.00757	0.00789	0.00809	0.00849
0.0033	0.384	0.00913	0.00932	0.0097	0.00984	0.01035
0.0031	0.334	0.01181	0.01204	0.01265	0.01291	0.01353
0.00145	0.29	0.00753	0.00784	0.00815	0.00849	0.00874

Table A.19: Energy balance during the pressurization phase of the self-pressurization experiments with respect to the control volume shown in figure 4.10.

p_0 [kPa]	120	130	140	150	160
$dH_{e,L}$ [kJ]	6.84	11.17	15.9	18.47	20.02
$V_L\,dp$ [kJ]	0.29	0.44	0.58	0.73	0.89
$m_{\mathrm{evap}}\,h_{e,L,\mathrm{evap}}$ [kJ]	-0.91	-1.31	-1.68	-2.3	-2.74
$m_{\mathrm{evap}}\,h_v$ [kJ]	1.49	2.14	2.75	3.76	4.48
$dH_{e,U}$ [kJ]	1.51	2.29	2.99	3.78	4.6
$V_U\,dp$ [kJ]	0.43	0.66	0.85	1.08	1.31
$m_{\mathrm{evap}}\,h_{e,U,\mathrm{evap}}$ [kJ]	0.58	0.83	1.07	1.46	1.74
dQ_L [kJ]	7.13	11.55	16.39	19.2	20.87
dQ_U [kJ]	0.5	0.81	1.06	1.24	1.55
dQ_{tot} [kJ]	7.62	12.36	17.46	20.44	22.42

Table A.20: Energy balance during the sloshing phase of the self-pressurization experiments with respect to the control volume shown in figure 4.10.

p_0 [kPa]	120	130	140	150	160
$dH_{e,L}$ [kJ]	5.03	8.02	10.19	14.36	18.18
$V_L\,dp$ [kJ]	-0.08	-0.15	-0.23	-0.3	-0.38
$m_{\mathrm{condens}}\,h_{e,L,\mathrm{condens}}$ [kJ]	-0.35	-0.38	-0.63	-0.81	-1.01
$m_{\mathrm{condens}}\,h_v$ [kJ]	0.57	0.62	1.04	1.34	1.66
$dH_{e,U}$ [kJ]	-0.41	-0.79	-1.18	-1.53	-1.98
$V_U\,dp$ [kJ]	-0.12	-0.22	-0.34	-0.44	-0.57
$m_{\mathrm{condens}}\,h_{e,U,\mathrm{condens}}$ [kJ]	0.22	0.24	0.41	0.53	0.65
dQ_L [kJ]	4.89	7.93	10.01	14.13	17.91
dQ_U [kJ]	-0.07	-0.32	-0.44	-0.57	-0.76
dQ_{tot} [kJ]	4.82	7.62	9.57	13.56	17.15

Table A.21: Tabular data of the liquid temperature sensors during GN_2 pressurization experiments. Please note, that the data provided for $z_{pos} = 0.29$ m (at the free surface) is not measured but determined based on equation (2.82).

initial pressure level	$p_0 = 120$ kPa	130	140	150	160
start press	$p = 101.28$ kPa	101.27	101.34	101.4	101.35
z_{pos} [m]		ϑ_L [K]			
0.29	77.35	77.35	77.36	77.36	77.36
0.278	77.49	77.62	77.64	77.61	77.67
0.253	77.5	77.6	77.68	77.62	77.67
0.203	77.64	77.77	77.8	77.76	77.81
0.103	77.54	77.64	77.75	77.66	77.72
0.003	77.53	77.64	77.69	77.63	77.68
end press	$p = 120$ kPa	130	140	150	160
0.29	78.82	79.53	80.21	80.84	81.45
0.278	77.58	77.75	77.86	77.84	78.16
0.253	77.55	77.68	77.74	77.69	77.74
0.203	77.67	77.81	77.86	77.8	77.86
0.103	77.54	77.66	77.73	77.68	77.73
0.003	77.55	77.67	77.72	77.66	77.71
end slosh	$p = 112.98$ kPa	114.81	117.57	120.94	123.48
0.29	78.29	78.43	78.64	78.89	79.07
0.278	78.18	78.33	78.53	78.77	78.98
0.253	78.13	78.23	78.5	78.7	78.88
0.203	77.77	78.07	78.14	78.09	78.27
0.103	77.57	77.7	77.81	77.69	77.76
0.003	77.54	77.68	77.76	77.68	77.73

Table A.22: Tabular data of the liquid density during GN_2 pressurization experiments taken from [41].

initial pressure level	$p_0 = 120$ kPa	130	140	150	160
start press	$p = 101.28$ kPa	101.27	101.34	101.4	101.35
z_{pos} [m]		$\varrho_L\ [\mathrm{kg\,m^{-3}}]$			
0.29	806.1	806.11	806.08	806.06	806.07
0.278	805.46	804.89	804.79	804.93	804.68
0.253	805.44	804.96	804.62	804.89	804.67
0.203	804.79	804.22	804.06	804.24	804.02
0.103	805.25	804.79	804.28	804.68	804.43
0.003	805.29	804.8	804.56	804.83	804.61
end press	$p = 120$ kPa	130	140	150	160
0.29	799.39	796.1	792.98	790	787.15
0.278	805.11	804.36	803.85	804	802.57
0.253	805.26	804.66	804.43	804.67	804.46
0.203	804.71	804.09	803.89	804.19	803.94
0.103	805.28	804.78	804.49	804.74	804.52
0.003	805.23	804.74	804.53	804.84	804.63
end slosh	$p = 112.98$ kPa	114.81	117.57	120.94	123.48
0.29	801.82	801.18	800.22	799.08	798.23
0.278	802.34	801.64	800.71	799.64	798.67
0.253	802.55	802.09	800.87	799.95	799.15
0.203	804.21	802.83	802.54	802.79	801.97
0.103	805.14	804.53	804.03	804.58	804.31
0.003	805.28	804.64	804.26	804.67	804.44

Table A.23: Tabular data of the liquid enthalpy during GN_2 pressurization experiments taken from [41].

initial pressure level	$p_0 = 120$ kPa	130	140	150	160
start press	$p = 101.28$ kPa	101.27	101.34	101.4	101.35
z_{pos} [m]	$h_{e,L}$ $[\text{J kg}^{-1}]$				
0.29	-122025.9	-122027.6	-122015.6	-122005.3	-122013.9
0.278	-121735.3	-121476.5	-121433.6	-121499.3	-121382.2
0.253	-121724.7	-121509.2	-121355.6	-121480.9	-121378.1
0.203	-121433.6	-121175.1	-121103.3	-121187.4	-121086.9
0.103	-121638.6	-121431.6	-121203.7	-121384.2	-121269.5
0.003	-121657.0	-121439.7	-121331.1	-121452.0	-121351.5
end press	$p = 120$ kPa	130	140	150	160
0.29	-119010.6	-117539.8	-116149.3	-114829.3	-113571.5
0.278	-121547.8	-121195.7	-120951.7	-120995.9	-120341.0
0.253	-121617.2	-121330.5	-121207.1	-121296.2	-121185.1
0.203	-121370.1	-121077.1	-120966.0	-121083.7	-120952.2
0.103	-121625.4	-121383.6	-121233.7	-121328.9	-121209.6
0.003	-121603.0	-121365.2	-121252.1	-121371.8	-121260.7
end slosh	$p = 112.98$ kPa	114.81	117.57	120.94	123.48
0.29	-120097.8	-119809.7	-119381.5	-118868.6	-118488.9
0.278	-120328.3	-120014.0	-119598.7	-119117.2	-118683.1
0.253	-120422.4	-120214.6	-119668.3	-119256.4	-118894.1
0.203	-121158.4	-120543.9	-120409.0	-120513.1	-120143.1
0.103	-121573.1	-121298.1	-121071.4	-121312.3	-121183.8
0.003	-121638.5	-121347.1	-121173.6	-121351.1	-121245.1

Table A.24: Tabular data of the liquid mass during GN_2 pressurization experiments.

initial pressure level		$p_0 = 120$ kPa	130	140	150	160
start press		$p = 101.28$ kPa	101.27	101.34	101.4	101.35
$V_{L,k}$ [m³]	z_{pos} [m]		$m_{L,k}$ [kg]			
0.0004	0.29	0.3195	0.3195	0.3195	0.3194	0.3195
0.00122	0.278	0.9842	0.9835	0.9834	0.9836	0.9833
0.00248	0.253	1.995	1.9938	1.993	1.9937	1.9931
0.00762	0.203	6.1309	6.1265	6.1253	6.1267	6.125
0.0026	0.103	2.0921	2.0909	2.0896	2.0906	2.09
0.00112	0.003	0.9049	0.9043	0.9041	0.9044	0.9041
end press		$p = 120$ kPa	130	140	150	160
0.0004	0.29	0.3168	0.3155	0.3143	0.3131	0.312
0.00122	0.278	0.9838	0.9829	0.9823	0.9825	0.9807
0.00248	0.253	1.9946	1.9931	1.9925	1.9931	1.9926
0.00762	0.203	6.1303	6.1256	6.124	6.1263	6.1244
0.0026	0.103	2.0922	2.0909	2.0901	2.0908	2.0902
0.00112	0.003	0.9048	0.9043	0.904	0.9044	0.9042
end slosh		$p = 112.98$ kPa	114.81	117.57	120.94	123.48
0.0004	0.29	0.3178	0.3175	0.3171	0.3167	0.3163
0.00122	0.278	0.9804	0.9796	0.9784	0.9771	0.9759
0.00248	0.253	1.9879	1.9867	1.9837	1.9814	1.9794
0.00762	0.203	6.1264	6.1159	6.1137	6.1156	6.1094
0.0026	0.103	2.0918	2.0902	2.0889	2.0904	2.0896
0.00112	0.003	0.9049	0.9042	0.9037	0.9042	0.9039

Table A.25: Tabular data of the ullage temperature sensors during GN_2 pressurization experiments. Please note, that the data provided for $z_{pos} = 0.29$ m (at the free surface) is not measured but determined based on equation (2.82).

initial pressure level	$p_0 = 120$ kPa	130	140	150	160
start press	$p = 101.28$ kPa	101.27	101.34	101.4	101.35
z_{pos} [m]		ϑ_U [K]			
0.635	290.00	288.00	285.00	287.00	286.00
0.484	185.38	187.08	184.66	188.70	184.73
0.434	161.12	160.10	157.62	160.71	157.78
0.384	139.45	132.48	129.96	134.27	130.58
0.334	117.02	109.04	107.66	111.62	108.51
0.29	77.35	77.35	77.36	77.36	77.36
end press	$p = 120$ kPa	130	140	150	160
0.635	290.00	288.00	285.00	287.00	286.00
0.484	192.49	195.15	193.25	198.32	194.44
0.434	166.10	167.30	165.77	169.36	167.10
0.384	143.82	139.34	137.75	142.04	139.59
0.334	121.25	115.07	114.44	118.53	116.51
0.29	78.82	79.53	80.21	80.84	81.45
end slosh	$p = 112.98$ kPa	114.81	117.57	120.94	123.48
0.635	290.00	288.00	285.00	287.00	286.00
0.484	190.63	192.37	190.27	195.33	191.66
0.434	164.55	164.97	163.41	167.29	165.18
0.384	139.11	135.89	134.55	138.38	136.09
0.334	104.47	101.71	101.90	102.60	102.96
0.29	78.29	78.43	78.64	78.89	79.07

Table A.26: Tabular data of the ullage density during GN_2 pressurization experiments taken from [41].

initial pressure level	$p_0 = 120$ kPa	130	140	150	160
start press	$p = 101.28$ kPa	101.27	101.34	101.4	101.35
z_{pos} [m]	ϱ_U [kg m^{-3}]				
0.635	1.18	1.19	1.2	1.19	1.19
0.484	1.85	1.83	1.85	1.82	1.85
0.434	2.13	2.14	2.18	2.14	2.18
0.384	2.47	2.6	2.65	2.57	2.64
0.334	2.95	3.18	3.22	3.11	3.2
0.29	4.61	4.61	4.61	4.62	4.61
end press	$p = 120$ kPa	130	140	150	160
0.635	1.39	1.52	1.66	1.76	1.89
0.484	2.11	2.25	2.45	2.56	2.78
0.434	2.45	2.63	2.86	3	3.25
0.384	2.83	3.17	3.46	3.6	3.91
0.334	3.38	3.87	4.2	4.34	4.72
0.29	5.39	5.81	6.22	6.63	7.04
end slosh	$p = 112.98$ kPa	114.81	117.57	120.94	123.48
0.635	1.31	1.34	1.39	1.42	1.46
0.484	2	2.02	2.09	2.09	2.18
0.434	2.32	2.36	2.44	2.45	2.53
0.384	2.76	2.87	2.97	2.97	3.09
0.334	3.72	3.89	3.97	4.06	4.13
0.29	5.1	5.18	5.29	5.43	5.54

Table A.27: Tabular data of the ullage enthalpy during GN_2 pressurization experiments taken from [41].

initial pressure level	$p_0 = 120$ kPa	130	140	150	160
start press	$p = 101.28$ kPa	101.27	101.34	101.4	101.35
z_pos [m]		$h_{e,U}$ $[\mathrm{J\,kg^{-1}}]$			
0.635	300780.3	298697.6	295573.4	297655.9	296614.7
0.484	191729.3	193510.1	190982.2	195195.8	191055.3
0.434	166364.6	165296.8	162699.4	165934.6	162866.9
0.384	143646.8	136317	133657.5	138194.1	134310.4
0.334	120000.7	111533.5	110063.9	114268.3	110968.5
0.29	77155.0	77154.4	77158.8	77162.6	77159.5
end press	$p = 120$ kPa	130	140	150	160
0.635	300735.8	298628.3	295478.1	297537.9	296471.3
0.484	199055.6	201783.6	199745.7	204997.1	200892.5
0.434	171449.1	172639	170966.8	174671.5	172232.4
0.384	148072	143259.1	141492.8	145924.3	143243.9
0.334	124251.1	117551	116750.7	120984.3	118696
0.29	78250.1	78767.0	79244.9	79688.7	80102.5
end slosh	$p = 112.98$ kPa	114.81	117.57	120.94	123.48
0.635	300752.5	298665	295533.4	297608.5	296560.6
0.484	197149.4	198957.4	196754	202017.9	198170
0.434	169879	170306.2	168652.8	172694.6	170461.4
0.384	143180.6	139773.3	138334.2	142332.3	139894.6
0.334	106473.6	103488.8	103641.4	104343.5	104688.9
0.29	77860.7	77964.5	78117.9	78300.5	78434.8

Table A.28: Tabular data of the ullage mass during GN_2 pressurization experiments.

initial pressure level		$p_0 = 120$ kPa	130	140	150	160
start press		$p = 101.28$ kPa	101.27	101.34	101.4	101.35
$V_{U,k}$ [m³]	z_{pos} [m]			$m_{U,k}$ [kg]		
0.00832	0.635	0.0098	0.00986	0.00997	0.00991	0.00994
0.0033	0.484	0.0061	0.00604	0.00612	0.006	0.00612
0.0033	0.434	0.00703	0.00707	0.00719	0.00705	0.00718
0.0033	0.384	0.00814	0.00858	0.00876	0.00847	0.00872
0.0031	0.334	0.00917	0.00987	0.01001	0.00964	0.00992
0.00145	0.29	0.0067	0.0067	0.0067	0.00671	0.0067
end press		$p = 120$ kPa	130	140	150	160
0.00832	0.635	0.01161	0.01266	0.01378	0.01466	0.01569
0.0033	0.484	0.00696	0.00744	0.00809	0.00844	0.00919
0.0033	0.434	0.00808	0.00869	0.00945	0.00991	0.01072
0.0033	0.384	0.00936	0.01048	0.01143	0.01187	0.0129
0.0031	0.334	0.01049	0.01202	0.01304	0.01348	0.01466
0.00145	0.29	0.00783	0.00844	0.00904	0.00963	0.01023
end slosh		$p = 112.98$ kPa	114.81	117.57	120.94	123.48
0.00832	0.635	0.01093	0.01118	0.01157	0.01182	0.01211
0.0033	0.484	0.00661	0.00666	0.0069	0.00691	0.00719
0.0033	0.434	0.00768	0.00778	0.00805	0.00808	0.00836
0.0033	0.384	0.00911	0.00949	0.00982	0.00981	0.0102
0.0031	0.334	0.01154	0.01207	0.01234	0.01261	0.01283
0.00145	0.29	0.00741	0.00752	0.00769	0.00789	0.00804

Table A.29: Energy balance during the pressurization phase of the GN_2 pressurization experiments with respect to the control volume shown in figure 4.19.

p_0 [kPa]	120	130	140	150	160
$dH_{e,L}$ [kJ]	2.32	3.59	4.39	4.94	6.4
$V_L\,dp$ [kJ]	0.29	0.44	0.6	0.75	0.91
$m_{\text{condens}}\,h_{e,L,\text{condens}}$ [kJ]	-0.02	-0.2	-0.35	-0.45	-0.74
$m_{\text{condens}}\,h_v$ [kJ]	0.03	0.32	0.56	0.74	1.21
$dH_{e,U}$ [kJ]	1.49	2.29	3.08	3.88	4.68
$V_U\,dp$ [kJ]	0.43	0.65	0.88	1.11	1.34
$m_{\text{GN2}}\,h_{e,\text{GN2}}$ [kJ]	2.3	4.02	5.74	7.28	9.38
$m_{\text{condens}}\,h_{e,U,\text{condens}}$ [kJ]	0.01	0.13	0.22	0.29	0.47
dQ_L [kJ]	2.02	3.02	3.57	3.91	5.03
dQ_U [kJ]	-1.22	-2.26	-3.32	-4.22	-5.57
dQ_{tot} [kJ]	0.8	0.76	0.25	-0.32	-0.54

Table A.30: Energy balance during the sloshing phase of the GN_2 pressurization experiments with respect to the control volume shown in figure 4.19.

p_0 [kPa]	120	130	140	150	160
$dH_{e,L}$ [kJ]	6.22	8.31	9.79	11.16	13.19
$V_L\,dp$ [kJ]	-0.11	-0.23	-0.35	-0.45	-0.56
$m_{\text{condens}}\,h_{e,L,\text{condens}}$ [kJ]	-0.13	-0.6	-1.01	-1.29	-1.74
$m_{\text{condens}}\,h_v$ [kJ]	0.21	0.99	1.67	2.14	2.89
$dH_{e,U}$ [kJ]	-0.56	-1.21	-1.79	-2.32	-2.91
$V_U\,dp$ [kJ]	-0.16	-0.35	-0.51	-0.66	-0.83
$m_{\text{condens}}\,h_{e,U,\text{condens}}$ [kJ]	0.08	0.39	0.66	0.85	1.15
dQ_L [kJ]	4.89	7.93	10.01	14.13	17.91
dQ_U [kJ]	-0.32	-0.47	-0.62	-0.8	-0.93
dQ_{tot} [kJ]	5.93	7.68	8.86	9.95	11.67

Table A.31: Tabular data of the liquid temperature sensors during GHe pressurization experiments. Please note, that the data provided for $z_{pos} = 0.29$ m (at the free surface) is not measured but determined based on equation (2.82).

GHe concentration	$\chi_{GHe, f+p} = 0.38$	0.46	0.52	0.6	0.68	0.73
start flush	$p = 105.63$ kPa	105.63	105.62	105.63	105.63	105.63
z_{pos} [m]			ϑ_L [K]			
0.29	77.71	77.71	77.71	77.71	77.71	77.71
0.278	77.37	77.6	77.39	77.53	77.43	77.39
0.253	77.38	77.61	77.41	77.53	77.43	77.44
0.203	77.50	77.75	77.53	77.66	77.56	77.57
0.103	77.40	77.67	77.41	77.54	77.44	77.44
0.003	77.36	77.65	77.40	77.54	77.44	77.45
end flush	$p = 102.1$ kPa	102.1	101.4	102.1	102.2	101.7
0.29	77.42	77.42	77.36	77.42	77.43	77.39
0.278	77.37	77.62	77.39	77.49	77.39	77.32
0.253	77.37	77.63	77.41	77.48	77.43	77.38
0.203	77.50	77.76	77.54	77.65	77.56	77.55
0.103	77.39	77.67	77.41	77.53	77.44	77.41
0.003	77.38	77.64	77.40	77.52	77.42	77.36
end press	$p = 140$ kPa	140	140	140	140	140
0.29	79.69	79.36	79.00	78.64	78.30	77.95
0.278	77.48	77.74	77.49	77.72	77.52	77.43
0.253	77.42	77.68	77.46	77.58	77.45	77.43
0.203	77.53	77.78	77.57	77.67	77.56	77.55
0.103	77.41	77.67	77.42	77.59	77.44	77.39
0.003	77.40	77.68	77.43	77.58	77.44	77.40
end slosh	$p = 124.46$ kPa	127.6	132.58	135.06	138.41	140.41
0.29	77.94	78.17	77.79	77.90	77.69	77.55
0.278	77.94	78.17	77.79	77.90	77.69	77.55
0.253	77.85	78.04	77.79	77.78	77.62	77.49
0.203	77.55	77.79	77.57	77.71	77.58	77.54
0.103	77.40	77.65	77.45	77.55	77.44	77.40
0.003	77.40	77.66	77.42	77.54	77.43	77.39

Table A.32: Tabular data of the liquid density during GHe pressurization experiments taken from [41].

GHe concentration	$\chi_{GHe,f+p} = 0.38$	0.46	0.52	0.6	0.68	0.73
start flush	$p = 105.63$ kPa	105.63	105.62	105.63	105.63	105.63
z_{pos} [m]	ϱ_L [kg m^{-3}]					
0.29	804.47	804.47	804.47	804.47	804.47	804.47
0.278	806.02	804.96	805.93	805.28	805.74	805.95
0.253	805.97	804.93	805.83	805.27	805.74	805.70
0.203	805.41	804.28	805.28	804.69	805.15	805.10
0.103	805.90	804.64	805.83	805.25	805.69	805.69
0.003	806.07	804.75	805.88	805.25	805.72	805.68
end flush	$p = 102.1$ kPa	102.1	101.4	102.1	102.2	101.7
0.29	805.79	805.79	806.06	805.79	805.75	805.94
0.278	806.03	804.89	805.92	805.45	805.93	806.24
0.253	806.00	804.84	805.82	805.51	805.76	805.99
0.203	805.42	804.27	805.25	804.75	805.16	805.19
0.103	805.92	804.67	805.82	805.27	805.72	805.84
0.003	805.97	804.77	805.86	805.33	805.79	806.07
end press	$p = 140$ kPa	140	140	140	140	140
0.29	795.40	796.94	798.61	800.28	801.87	803.48
0.278	805.61	804.45	805.57	804.52	805.44	805.83
0.253	805.88	804.71	805.72	805.15	805.76	805.83
0.203	805.38	804.22	805.22	804.73	805.26	805.32
0.103	805.92	804.73	805.87	805.13	805.80	806.02
0.003	806.00	804.68	805.87	805.14	805.80	805.97
end slosh	$p = 124.46$ kPa	127.6	132.58	135.06	138.41	140.41
0.29	803.45	802.44	804.17	803.68	804.67	805.29
0.278	803.45	802.44	804.17	803.68	804.67	805.29
0.253	803.87	802.99	804.18	804.24	804.98	805.57
0.203	805.25	804.17	805.16	804.56	805.14	805.34
0.103	805.93	804.79	805.72	805.28	805.79	805.98
0.003	805.96	804.77	805.89	805.32	805.86	806.02

Table A.33: Tabular data of the liquid enthalpy during GHe pressurization experiments taken from [41].

GHe conc. $\chi_{GHe,f+p}$	0.38	0.46	0.52	0.6	0.68	0.73
start flush	$p = 105.63$ kPa	105.63	105.62	105.63	105.63	105.63
z_{pos} [m]		$h_{e,L}$ [J kg^{-1}]				
0.29	-121291	-121290	-121292	-121290	-121290	-121290
0.278	-121981	-121510	-121942	-121653	-121858	-121953
0.253	-121961	-121495	-121899	-121651	-121858	-121841
0.203	-121711	-121204	-121652	-121386	-121593	-121575
0.103	-121928	-121364	-121897	-121638	-121837	-121839
0.003	-122003	-121417	-121928	-121643	-121846	-121833
end flush	$p = 102.1$ kPa	102.1	101.4	102.1	102.2	101.7
0.29	-121886	-121887	-122005	-121887	-121869	-121954
0.278	-121996	-121481	-121944	-121733	-121949	-122088
0.253	-121979	-121458	-121897	-121760	-121870	-121975
0.203	-121719	-121198	-121648	-121417	-121602	-121614
0.103	-121942	-121378	-121897	-121651	-121852	-121910
0.003	-121965	-121425	-121917	-121675	-121882	-122012
end press	$p = 140$ kPa	140	140	140	140	140
0.29	-117211	-117888	-118627	-119365	-120070	-120785
0.278	-121734	-121215	-121716	-121248	-121656	-121832
0.253	-121855	-121332	-121785	-121528	-121799	-121832
0.203	-121634	-121115	-121563	-121342	-121577	-121603
0.103	-121873	-121340	-121852	-121522	-121818	-121920
0.003	-121910	-121320	-121848	-121524	-121818	-121897
end slosh	$p = 124.46$ kPa	127.6	132.58	135.06	138.41	140.41
0.29	-120799	-120345	-121106	-120884	-121317	-121591
0.278	-120799	-120345	-121106	-120884	-121317	-121591
0.253	-120987	-120592	-121112	-121131	-121457	-121717
0.203	-121604	-121118	-121549	-121274	-121527	-121611
0.103	-121904	-121393	-121798	-121595	-121817	-121901
0.003	-121920	-121381	-121872	-121613	-121847	-121915

Table A.34: Tabular data of the liquid mass during GHe pressurization experiments.

GHe conc. $\chi_{GHe,f+p}$		0.38	0.46	0.52	0.6	0.68	0.73
start flush		$p = 105.63$ kPa	105.63	105.62	105.63	105.63	105.63
$V_{L,k}$ [m^3]	z_{pos} [m]			$m_{L,k}$ [kg]			
0.0004	0.29	0.3188	0.3188	0.3188	0.3188	0.3188	0.3188
0.00122	0.278	0.9849	0.9836	0.9848	0.9840	0.9846	0.9848
0.00248	0.253	1.9963	1.9938	1.9960	1.9946	1.9958	1.9957
0.00762	0.203	6.1356	6.1270	6.1346	6.1301	6.1336	6.1332
0.0026	0.103	2.0938	2.0905	2.0936	2.0921	2.0932	2.0933
0.00112	0.003	0.9058	0.9043	0.9056	0.9049	0.9054	0.9053
end flush		$p = 102.1$ kPa	102.1	101.4	102.1	102.2	101.7
0.0004	0.29	0.3193	0.3193	0.3194	0.3193	0.3193	0.3194
0.00122	0.278	0.9849	0.9835	0.9848	0.9842	0.9848	0.9852
0.00248	0.253	1.9964	1.9936	1.9960	1.9952	1.9958	1.9964
0.00762	0.203	6.1357	6.1269	6.1344	6.1306	6.1337	6.1339
0.0026	0.103	2.0938	2.0906	2.0936	2.0922	2.0933	2.0936
0.00112	0.003	0.9057	0.9043	0.9055	0.9049	0.9054	0.9058
end press		$p = 140$ kPa	140	140	140	140	140
0.0004	0.29	0.3152	0.3158	0.3165	0.3172	0.3178	0.3184
0.00122	0.278	0.9844	0.9830	0.9844	0.9831	0.9842	0.9847
0.00248	0.253	1.9961	1.9932	1.9957	1.9943	1.9958	1.9960
0.00762	0.203	6.1354	6.1265	6.1342	6.1304	6.1344	6.1349
0.0026	0.103	2.0938	2.0907	2.0937	2.0918	2.0935	2.0941
0.00112	0.003	0.9057	0.9042	0.9055	0.9047	0.9055	0.9057
end slosh		$p = 124.46$ kPa	127.6	132.58	135.06	138.41	140.41
0.0004	0.29	0.3184	0.3180	0.3187	0.3185	0.3189	0.3191
0.00122	0.278	0.9818	0.9805	0.9827	0.9821	0.9833	0.9840
0.00248	0.253	1.9911	1.9890	1.9919	1.9921	1.9939	1.9954
0.00762	0.203	6.1344	6.1262	6.1337	6.1291	6.1335	6.1350
0.0026	0.103	2.0939	2.0909	2.0933	2.0922	2.0935	2.0940
0.00112	0.003	0.9056	0.9043	0.9056	0.9049	0.9055	0.9057

Table A.35: Tabular data of the ullage temperature sensors during GHe pressurization experiments. Please note, that the data provided for $z_\mathrm{pos} = 0.29$ m (at the free surface) is not measured but determined based on equation (2.82).

GHe concentration	$\chi_{GHe,f+p} = 0.38$	0.46	0.52	0.6	0.68	0.73
start flush	$p = 105.63$ kPa	105.63	105.62	105.63	105.63	105.63
z_pos [m]				ϑ_U [K]		
0.635	282.75	284.00	283.50	282.90	281.80	281.50
0.484	189.24	186.41	188.42	185.12	186.40	184.82
0.434	161.84	158.92	161.46	158.12	159.39	158.46
0.384	133.42	130.42	134.83	130.07	130.57	131.18
0.334	108.09	107.30	110.31	106.54	106.21	106.93
0.29	77.71	77.71	77.71	77.71	77.71	77.71
end flush	$p = 102.1$ kPa	102.1	101.4	102.1	102.2	101.7
0.635	282.75	284.00	283.50	282.90	281.80	281.50
0.484	193.07	192.70	196.27	194.19	196.09	195.03
0.434	166.10	166.38	170.44	169.46	172.18	172.02
0.384	138.05	138.59	144.47	142.85	146.04	147.85
0.334	113.59	115.65	121.33	120.22	123.81	127.65
0.29	77.42	77.42	77.36	77.42	77.43	77.39
end press	$p = 140$ kPa	140	140	140	140	140
0.635	282.75	284.00	283.50	282.90	281.80	281.50
0.484	198.90	199.19	202.23	199.02	200.32	198.83
0.434	173.26	173.55	177.31	175.29	177.32	176.46
0.384	144.08	145.16	150.79	148.93	151.73	152.38
0.334	116.54	119.60	125.03	124.28	126.80	129.11
0.29	80.21	80.21	80.21	80.21	80.21	80.21
end slosh	$p = 124.46$ kPa	127.60	132.58	135.06	138.41	140.41
0.635	282.75	284.00	283.50	282.90	281.80	281.50
0.484	197.49	197.96	199.94	197.65	198.84	197.93
0.434	170.25	169.62	174.10	172.45	175.23	173.98
0.384	138.25	138.90	143.60	142.69	144.96	146.05
0.334	103.60	105.85	106.73	106.52	106.11	108.43
0.29	79.14	79.37	79.71	79.88	80.10	80.23

Table A.36: Tabular data of the gaseous nitrogen density during GHe pressurization experiments taken from [41].

GHe concentration	$\chi_{GHe,f+p} = 0.38$	0.46	0.52	0.6	0.68	0.73
start flush	$p = 105.63$ kPa	105.63	105.62	105.63	105.63	105.63
z_{pos} [m]	ϱ_U [kg m^{-3}]					
0.635	1.26	1.25	1.26	1.26	1.26	1.26
0.484	1.89	1.91	1.89	1.93	1.92	1.93
0.434	2.21	2.25	2.21	2.26	2.24	2.26
0.384	2.69	2.75	2.66	2.76	2.75	2.74
0.334	3.35	3.37	3.28	3.40	3.41	3.39
0.29	4.79	4.79	4.79	4.79	4.79	4.79
end flush	$p = 102.1$ kPa	102.1	101.4	102.1	102.2	101.7
0.635	1.22	1.21	1.21	1.22	1.22	1.22
0.484	1.79	1.79	1.74	1.78	1.76	1.76
0.434	2.08	2.08	2.01	2.04	2.01	2.00
0.384	2.51	2.50	2.38	2.42	2.37	2.33
0.334	3.07	3.01	2.85	2.89	2.81	2.71
0.29	4.64	4.64	4.62	4.64	4.65	4.63
end press	$p = 140$ kPa	140	140	140	140	140
0.635	1.67	1.66	1.66	1.67	1.67	1.68
0.484	2.38	2.38	2.34	2.38	2.36	2.38
0.434	2.74	2.73	2.67	2.70	2.67	2.69
0.384	3.30	3.28	3.15	3.19	3.13	3.12
0.334	4.12	4.01	3.83	3.85	3.77	3.70
0.29	6.22	6.22	6.22	6.22	6.22	6.22
end slosh	$p = 124.46$ kPa	127.6	132.58	135.06	138.41	140.41
0.635	1.48	1.51	1.58	1.61	1.66	1.68
0.484	2.13	2.18	2.24	2.31	2.35	2.40
0.434	2.47	2.55	2.58	2.65	2.67	2.73
0.384	3.06	3.12	3.14	3.22	3.25	3.27
0.334	4.14	4.15	4.28	4.37	4.50	4.46
0.29	5.58	5.71	5.91	6.01	6.15	6.24

Table A.37: Tabular data of the gaseous helium density during GHe pressurization experiments taken from [41].

GHe concentration	z_{pos} [m]	$\chi_{GHe,f+p} = 0.38$	0.46	0.52	0.6	0.68	0.73
start flush		$p = 105.63$ kPa	105.63	105.62	105.63	105.63	105.63
		ϱ_U [kg m^{-3}]					
end flush		$p = 102.1$ kPa	102.1	101.4	102.1	102.2	101.7
	0.635	0.17	0.17	0.17	0.17	0.17	0.17
	0.484	0.25	0.25	0.25	0.25	0.25	0.25
	0.434	0.30	0.30	0.29	0.29	0.29	0.28
	0.384	0.36	0.35	0.34	0.34	0.34	0.33
	0.334	0.43	0.42	0.40	0.41	0.40	0.38
	0.29	0.63	0.63	0.63	0.63	0.63	0.63
end press		$p = 140$ kPa	140	140	140	140	140
	0.635	0.24	0.24	0.24	0.24	0.24	0.24
	0.484	0.34	0.34	0.33	0.34	0.34	0.34
	0.434	0.39	0.39	0.38	0.38	0.38	0.38
	0.384	0.47	0.46	0.45	0.45	0.44	0.44
	0.334	0.58	0.56	0.54	0.54	0.53	0.52
	0.29	0.84	0.84	0.84	0.84	0.84	0.84
end slosh		$p = 124.46$ kPa	127.6	132.58	135.06	138.41	140.41
	0.635	0.21	0.22	0.22	0.23	0.24	0.24
	0.484	0.30	0.31	0.32	0.33	0.33	0.34
	0.434	0.35	0.36	0.37	0.38	0.38	0.39
	0.384	0.43	0.44	0.44	0.46	0.46	0.46
	0.334	0.58	0.58	0.60	0.61	0.63	0.62
	0.29	0.76	0.77	0.80	0.81	0.83	0.84

Table A.38: Tabular data of the gaseous nitrogen enthalpy during GHe pressurization experiments taken from [41].

GHe conc. $\chi_{GHe,f+p}$	0.38	0.46	0.52	0.6	0.68	0.73
start flush	$p = 105.63$ kPa	105.63	105.62	105.63	105.63	105.63
z_{pos} [m]	$h_{e,U}$ [J kg^{-1}]					
0.635	293220	294521	294001	293376	292230	291918
0.484	195737	192781	194881	191433	192771	191120
0.434	167087	164035	166690	163191	164527	163547
0.384	137263	134097	138742	133728	134255	134898
0.334	110453	109616	112817	108800	108454	109222
0.29	77426	77426	77426	77427	77426	77427
end flush	$p = 102.1$ kPa	102.1	101.4	102.1	102.2	101.7
0.635	293228	294530	294011	293385	292239	291928
0.484	199755	199374	203104	200924	202907	201803
0.434	171571	171864	176121	175086	177935	177766
0.384	142162	142735	148912	147205	150553	152462
0.334	116350	118539	124558	123376	127163	131225
0.29	77207	77207	77163	77207	77213	77182
end press	$p = 140$ kPa	140	140	140	140	140
0.635	293134	294436	293915	293290	292144	291831
0.484	205653	205961	209133	205784	207142	205580
0.434	178820	179129	183069	180953	183074	182173
0.384	148165	149303	155236	153273	156219	156908
0.334	118988	122255	128028	127226	129906	132355
0.29	79245	79245	79245	79245	79245	79245
end slosh	$p = 124.46$ kPa	127.6	132.58	135.06	138.41	140.41
0.635	293172	294467	293934	293302	292148	291830
0.484	204262	204737	206782	204376	205598	204637
0.434	175767	175086	179747	178008	180900	179577
0.384	142162	142818	147725	146744	149111	150237
0.334	105356	107729	108592	108328	107841	110300
0.29	78486	78647	78894	79013	79171	79264

Table A.39: Tabular data of the gaseous helium enthalpy during GHe pressurization experiments taken from [41].

GHe conc. $\chi_{GHe,f+p}$		0.38	0.46	0.52	0.6	0.68	0.73
start flush		$p = 105.63$ kPa	105.63	105.62	105.63	105.63	105.63
z_{pos} [m]				$h_{e,U}$ $[\mathrm{J\,kg^{-1}}]$			
end flush		$p = 102.1$ kPa	102.1	101.4	102.1	102.2	101.7
	0.635	1473882	1480373	1477775	1474661	1468949	1467390
	0.484	1008144	1006248	1024785	1013960	1023827	1018321
	0.434	868083	869537	890646	885532	899684	898826
	0.384	722409	725239	755749	747338	763905	773330
	0.334	595371	606096	635595	629832	648452	668420
	0.29	407505	407505	407199	407505	407548	407330
end press		$p = 140$ kPa	140	140	140	140	140
	0.635	1474005	1480496	1477900	1474784	1469072	1467514
	0.484	1038543	1040075	1055836	1039192	1045943	1038180
	0.434	905389	906921	926448	915958	926474	922008
	0.384	753846	759455	788720	779034	793576	796979
	0.334	610807	626727	654931	651009	664124	676122
	0.29	422079	422079	422079	422079	422079	4220792
end slosh		$p = 124.46$ kPa	127.6	132.58	135.06	138.41	140.41
	0.635	1473955	1480456	1477876	1474768	1469066	1467515
	0.484	1031196	1033647	1043946	1032062	1038226	1033507
	0.434	889707	886446	909728	901193	915641	909155.9
	0.384	723518	726904	751329	746611	758437	764078
	0.334	543547	555269	559829	558745	556651	568682
	0.29	416514	417679	419483	420363	421532	422220

Table A.40: Tabular data of the gaseous nitrogen mass during GHe pressurization experiments.

GHe conc. $\chi_{GHe,f+p}$		0.38	0.46	0.52	0.6	0.68	0.73
start flush		$p = 105.63$ kPa	105.63	105.62	105.63	105.63	105.63
$V_{L,k}$ [m^3]	z_{pos} [m]			$m_{U,k}$ [kg]			
0.00832	0.635	0.01048	0.01043	0.01045	0.01047	0.01051	0.01053
0.0033	0.484	0.00623	0.00632	0.00626	0.00637	0.00632	0.00638
0.0033	0.434	0.00730	0.00743	0.00731	0.00747	0.00741	0.00746
0.0033	0.384	0.00889	0.00910	0.00879	0.00912	0.00909	0.00904
0.0031	0.334	0.01039	0.01047	0.01017	0.01055	0.01059	0.01051
0.00145	0.29	0.00696	0.00696	0.00696	0.00696	0.00696	0.00696
end flush		$p = 102.1$ kPa	102.1	101.4	102.1	102.2	101.7
0.00832	0.635	0.01003	0.00979	0.00951	0.00930	0.00896	0.00841
0.0033	0.484	0.00584	0.00574	0.00546	0.00539	0.00512	0.00483
0.0033	0.434	0.00680	0.00666	0.00630	0.00618	0.00584	0.00548
0.0033	0.384	0.00821	0.00802	0.00745	0.00736	0.00690	0.00639
0.0031	0.334	0.00944	0.00909	0.00838	0.00826	0.00768	0.00699
0.00145	0.29	0.00668	0.00655	0.00636	0.00620	0.00595	0.00558
end press		$p = 140$ kPa	140	140	140	140	140
0.00832	0.635	0.01276	0.01235	0.01197	0.01145	0.01072	0.01003
0.0033	0.484	0.00722	0.00701	0.00668	0.00648	0.00600	0.00565
0.0033	0.434	0.00830	0.00806	0.00763	0.00737	0.00679	0.00638
0.0033	0.384	0.01002	0.00967	0.00900	0.00870	0.00796	0.00740
0.0031	0.334	0.01175	0.01112	0.01027	0.00986	0.00900	0.00826
0.00145	0.29	0.00830	0.00807	0.00781	0.00745	0.00695	0.00649
end slosh		$p = 124.46$ kPa	127.6	132.58	135.06	138.41	140.41
0.00832	0.635	0.01134	0.01126	0.01134	0.01105	0.01060	0.01006
0.0033	0.484	0.00646	0.00643	0.00639	0.00629	0.00598	0.00569
0.0033	0.434	0.00751	0.00752	0.00736	0.00723	0.00679	0.00649
0.0033	0.384	0.00929	0.00922	0.00896	0.00877	0.00824	0.00776
0.0031	0.334	0.01180	0.01151	0.01148	0.01119	0.01074	0.00995
0.00145	0.29	0.00744	0.00741	0.00742	0.00721	0.00688	0.00651

Table A.41: Tabular data of the gaseous helium mass during GHe pressurization experiments.

GHe conc. $\chi_{GHe,f+p}$		0.38	0.46	0.52	0.6	0.68	0.73
start flush		$p = 105.63$ kPa	105.63	105.62	105.63	105.63	105.63
$V_{L,i}$ [m³]	z_{pos} [m]	$m_{U,i}$ [kg]					
end flush		$p = 102.1$ kPa	102.1	101.4	102.1	102.2	101.7
0.00832	0.635	0.000014	0.000042	0.000074	0.000118	0.000174	0.000246
0.0033	0.484	0.000008	0.000024	0.000043	0.000068	0.000099	0.000141
0.0033	0.434	0.000010	0.000028	0.000049	0.000078	0.000113	0.000160
0.0033	0.384	0.000011	0.000034	0.000058	0.000092	0.000133	0.000186
0.0031	0.334	0.000013	0.000038	0.000065	0.000103	0.000147	0.000202
0.00145	0.29	0.000009	0.000027	0.000048	0.000075	0.000110	0.000156
end press		$p = 140$ kPa	140	140	140	140	140
0.00832	0.635	0.000161	0.000210	0.000269	0.000347	0.000459	0.000560
0.0033	0.484	0.000091	0.000119	0.000149	0.000196	0.000256	0.000315
0.0033	0.434	0.000104	0.000137	0.000170	0.000222	0.000290	0.000354
0.0033	0.384	0.000126	0.000163	0.000200	0.000261	0.000338	0.000410
0.0031	0.334	0.000146	0.000186	0.000227	0.000294	0.000380	0.000455
0.00145	0.29	0.000099	0.000130	0.000165	0.000213	0.000281	0.000343
end slosh		$p = 124.46$ kPa	127.6	132.58	135.06	138.41	140.41
0.00832	0.635	0.000143	0.000192	0.000254	0.000335	0.000454	0.000562
0.0033	0.484	0.000081	0.000109	0.000143	0.000190	0.000255	0.000317
0.0033	0.434	0.000094	0.000127	0.000164	0.000218	0.000290	0.000361
0.0033	0.384	0.000116	0.000156	0.000199	0.000263	0.000350	0.000429
0.0031	0.334	0.000146	0.000192	0.000252	0.000331	0.000449	0.000543
0.00145	0.29	0.000089	0.000120	0.000158	0.000206	0.000278	0.000344

Table A.42: Energy balance during the flushing phase of the GHe pressurization experiments with respect to the control volume shown in figure 4.30.

$\chi_{GHe,f+p}$ [mol mol^{-1}]	0.38	0.46	0.52	0.6	0.68	0.73
$dH_{e,L}$ [kJ]	-0.36	-0.12	-0.2	-0.97	-0.55	-1.54
$V_L\,dp$ [kJ]	-0.05	-0.05	-0.07	-0.05	-0.05	-0.06
$dH_{e,U}$ [kJ]	-0.30	-0.35	-0.46	-0.47	-0.55	-0.70
$V_U\,dp$ [kJ]	-0.08	-0.08	-0.10	-0.08	-0.08	-0.09
$m_{GHe}\,h_{e,GHe}$ [kJ]	0.58	1.56	2.52	3.50	4.47	5.57
$m_{GN2}\,h_{e,GN2}$ [kJ]	0.83	2.22	3.59	4.98	6.37	7.93
dQ_L [kJ]	-0.3	-0.06	-0.14	-0.92	-0.49	-1.48
dQ_U [kJ]	0.02	0.39	0.71	1.09	1.42	1.75
dQ_{tot} [kJ]	-0.28	0.33	0.57	0.17	0.93	0.26

Table A.43: Energy balance during the pressurization phase of the GHe pressurization experiments with respect to the control volume shown in figure 4.30.

$\chi_{GHe,f+p}$ [mol mol^{-1}]	0.38	0.46	0.52	0.6	0.68	0.73
$dH_{e,L}$ [kJ]	3.32	3.03	2.61	3.22	1.43	1.13
$V_L\,dp$ [kJ]	0.58	0.58	0.60	0.58	0.58	0.59
$m_{condens}\,h_{e,L,condens}$ [kJ]	-0.61	-0.55	-0.44	-0.59	-0.15	-0.07
$m_{condens}\,h_v$ [kJ]	1.02	0.92	0.73	0.99	0.24	0.11
$dH_{e,U}$ [kJ]	2.79	2.75	2.77	2.66	2.56	2.56
$V_U\,dp$ [kJ]	0.86	0.86	0.88	0.86	0.86	0.87
$m_{GHe}\,h_{e,GHe}$ [kJ]	4.11	4.02	3.89	3.82	3.82	3.53
$m_{condens}\,h_{e,U,condens}$ [kJ]	0.41	0.37	0.29	0.39	0.1	0.04
dQ_L [kJ]	2.33	2.07	1.73	2.24	0.75	0.49
dQ_U [kJ]	-1.77	-1.76	-1.71	-1.63	-2.02	-1.8
dQ_{tot} [kJ]	0.56	0.31	0.01	0.61	-1.27	-1.31

Table A.44: Energy balance during the sloshing phase of the GHe pressurization experiments with respect to the control volume shown in figure 4.30.

$\chi_{GHe,f+p}$ [mol mol^{-1}]	0.38	0.46	0.52	0.6	0.68	0.73
$dH_{e,L}$ [kJ]	2.29	1.93	1.83	1.18	1.22	0.24
$V_L\,dp$ [kJ]	-0.24	-0.19	-0.11	-0.08	-0.02	0.01
$m_{\text{condens}}\,h_{e,L,\text{condens}}$ [kJ]	-0.66	-0.55	-0.5	-0.32	-0.32	-0.06
$m_{\text{condens}}\,h_v$ [kJ]	1.09	0.91	0.83	0.54	0.52	0.1
$dH_{e,U}$ [kJ]	-2.48	-2.62	-2.70	-3.04	-3.48	-3.86
$V_U\,dp$ [kJ]	-0.35	-0.28	-0.17	-0.11	-0.04	0.01
$m_{\text{condens}}\,h_{e,U,\text{condens}}$ [kJ]	0.43	0.36	0.33	-0.21	-0.21	-0.04
dQ_L [kJ]	2.1	1.76	1.62	1.04	1.04	0.2
dQ_U [kJ]	-1.7	-1.98	-2.2	-2.72	-3.24	-3.83
dQ_{tot} [kJ]	0.4	-0.22	-0.58	-1.68	-2.2	-3.63

Table A.45: Properties for liquid nitrogen (LN$_2$) experiments for self-pressurization and GN$_2$ pressurization. The liquid density is $\varrho_L = 806.09$ kg m^{-3}, the ullage density is $\varrho_U = 4.61$ kg m^{-3}, the latent heat of evaporation is $\Delta h_v = 199$ kJ kg^{-1}, the heat capacity for constant pressure is the liquid is $c_{p,L} = 2041$ J kg^{-1} K^{-1}, the gravitational acceleration is $g = 9.81$ m s^{-2} and the tank radius is $R = 0.145$ m.

p_0	L_U	ϑ_{sat}	ϑ_{ref}	$\Theta_L = \vartheta_{\text{sat}} - \vartheta_{\text{ref}}$	**R Ja**	D_T [41]
[kPa]	[m]	[K]	[K]	[K]		[m^2/s]
120	0.37	78.82	77.35	1.47	2.63	8.71E-8
130	0.37	79.53	77.35	2.18	3.91	8.64E-8
140	0.37	80.21	77.35	2.86	5.13	8.57E-8
150	0.37	80.84	77.35	3.49	6.26	8.51E-8
160	0.37	81.45	77.35	4.10	7.35	8.44E-8

Table A.46: Pressure gradient data for liquid nitrogen (LN_2) for self-pressurization to apply the model by DAS & HOPFINGER [26]. Values for δ_t are determined by means of equation (2.91), u_v is determined by means of equation (2.84), D_e is determined by means of equation (2.93), D_e^* is determined by means of equation (2.94), D_e' is determined by means of equation (2.87) and $D_e'^*$ is determined by means of equation (2.95).

p_0	t_p	δ_t	$\frac{\partial p}{\partial t}$	$\frac{p_0}{\vartheta_0}\frac{\partial \vartheta_U}{\partial t}$	u_v	D_e	D_e^*	D_e'	$D_e'^*$
[kPa]	[s]	[m]	[Pa s^{-1}]	[Pa s^{-1}]	[m s^{-1}]	[m^2 s^{-1}]		[m^2 s^{-1}]	
120	740	0.024	-123	-37.50	-2.64E-4	2.41E-6	1.39E-5	2.32E-6	1.34E-5
130	1180	0.030	-180	-51.56	-3.66E-4	2.83E-6	1.64E-5	2.75E-6	1.59E-5
140	1800	0.037	-225	-62.48	-4.30E-4	3.12E-6	1.80E-5	3.03E-6	1.75E-5
150	2000	0.039	-246	-43.97	-4.98E-4	3.12E-6	1.80E-5	3.03E-6	1.75E-5
160	2432	0.043	-293	-47.34	-5.68E-4	3.32E-6	1.92E-5	3.24E-6	1.87E-5

Table A.47: Pressure gradient data for liquid nitrogen (LN_2) for GN_2 pressurization to apply the model by DAS & HOPFINGER [26]. Values for δ_t are determined by means of equation (2.91), u_v is determined by means of equation (2.84), D_e is determined by means of equation (2.93), D_e^* is determined by means of equation (2.94), D_e' is determined by means of equation (2.87) and $D_e'^*$ is determined by means of equation (2.95).

p_0	t_p	δ_t	$\frac{\partial p}{\partial t}$	$\frac{p_0}{\vartheta_0}\frac{\partial \vartheta_U}{\partial t}$	u_v	D_e	D_e^*	D_e'	$D_e'^*$
[kPa]	[s]	[m]	[Pa s^{-1}]	[Pa s^{-1}]	[m s^{-1}]	[m^2 s^{-1}]		[m^2 s^{-1}]	
120	72	0.0075	-115	-24.43	-2.79E-4	0.80E-6	4.6E-6	7.09E-7	4.10E-6
130	126	0.0099	-275	-59.92	-6.12E-4	1.55E-6	8.96E-6	1.46E-6	8.46E-6
140	180	0.012	-394	-57.65	-8.89E-4	2.04E-6	1.18E-5	1.96E-6	11.31E-6
150	228	0.013	-470	-53.08	-10.28E-4	2.17E-6	1.26E-5	2.09E-6	12.06E-6
160	294	0.015	-571	-49.03	-12.07E-4	2.45E-6	1.42E-5	2.37E-6	13.70E-6

Table A.48: Properties for liquid oxygen (LOX) experiments based on LACAPERE's results [39] provided by DAS & HOPFINGER [26]. The liquid density is $\varrho_L = 1091.89$ kg m^{-3} corresponding to an initial pressure of $p_0 = 250$ kPa, the ullage density is $\varrho_U = 10.27$ kg m^{-3}, the latent heat of evaporation is $\Delta h_v = 203$ kJ kg^{-1}, the heat capacity for constant pressure is the liquid is $c_{p,L} = 1737$ J kg^{-1} K^{-1}, the gravitational acceleration is $g = 9.81$ m s^{-2} and the tank radius is $R = 0.091$ m.

p_0	L_U	ϑ_{sat}	ϑ_{ref}	$\Theta_L = \vartheta_{\text{sat}} - \vartheta_{\text{ref}}$	$\mathbf{R\,Ja}$	D_T [41]
[kPa]	[m]	[K]	[K]	[K]		[m^2 s^{-1}]
250	0.37	99.81	90.00	9.81	8.93	7.34E-8

Table A.49: Modified pressure gradient data for liquid oxygen (LOX) for tests conducted by LACAPERE [39] to fit the current excitation of $\eta = 0.78$ by applying the model by DAS & HOPFINGER [26]. Values for δ_t are determined by means of equation (2.91). For known $D_e'^*$ taken from figure 4.36 for known δ_t, u_v is determined by means of equation (2.93), D_e is determined by means of equation (2.87), D_e^* is determined by means of equation (2.94), D_e' is determined by means of equation (2.95).

p_0	t_p	δ_t	$\frac{\partial p}{\partial t}$	$\frac{p_0}{\vartheta_0}\frac{\partial \vartheta_U}{\partial t}$	u_v	D_e	D_e^*	D_e'	$D_e'^*$
[kPa]	[s]	[m]	[Pa s^{-1}]	[Pa s^{-1}]	[m s^{-1}]	[m^2 s^{-1}]		[m^2 s^{-1}]	
250	87	0.0076	-1185	-104	-16.0E-4	0.68E-6	7.85E-6	6.02E-7	5.5E-6

Table A.50: Properties for liquid hydrogen (LH$_2$) based on MORAN's results [48] provided by DAS & HOPFINGER [26]. The liquid density is $\varrho_L = 66.35$ kg m^{-3} corresponding to an initial pressure of $p_0 = 250$ kPa, the ullage density is $\varrho_U = 3.08$ kg m^{-3}, the latent heat of evaporation is $\Delta h_v = 420$ kJ kg^{-1}, the heat capacity for constant pressure is the liquid is $c_{p,L} = 12300$ J kg^{-1} K^{-1}, the gravitational acceleration is $g = 9.81$ m s^{-2} and the tank radius is $R = 0.748$ m.

p_0	L_U	ϑ_{sat}	ϑ_{ref}	$\Theta_L = \vartheta_{\text{sat}} - \vartheta_{\text{ref}}$	$\mathbf{R\,Ja}$	D_T [41]
[kPa]	[m]	[K]	[K]	[K]		[m^2 s^{-1}]
239	0.49	23.76	20.26	3.5	2.21	1.28E-7
246	0.53	23.76	20.26	3.5	2.21	1.28E-7

Table A.51: Modified pressure gradient data for liquid hydrogen (LH$_2$) for tests conducted by MORAN [48] to fit the current excitation of $\eta = 0.78$ by applying the model by DAS & HOPFINGER [26]. Values for δ_t are determined by means of equation (2.91). For known $D_e'^*$ taken from figure 4.36 for known δ_t, u_v is determined by means of equation (2.93), D_e is determined by means of equation (2.87), D_e^* is determined by means of equation (2.94), D_e' is determined by means of equation (2.95).

p_0	t_p	δ_t	$\frac{\partial p}{\partial t}$	$\frac{p_0}{\vartheta_0}\frac{\partial \vartheta_U}{\partial t}$	u_v	D_e	D_e^*	D_e'	$D_e'^*$
[kPa]	[s]	[m]	[Pa s^{-1}]	[Pa s^{-1}]	[m s^{-1}]	[m^2 s^{-1}]		[m^2 s^{-1}]	
239	36	0.0064	-1140	-100	-2.13E-3	6.21E-6	3.06E-6	6.08E-6	3.00E-6
246	34	0.0063	-1084	-103	-2.19E-3	6.21E-6	3.06E-6	6.08E-6	3.00E-6

Table A.52: Properties for HFE7000 based on the results provided by DAS & HOPFINGER [26]. The liquid density is $\varrho_L = 1310$ kg m^{-3} corresponding to an initial pressure of $p_0 = 220$ kPa, the ullage density is $\varrho_U = 14.61$ kg m^{-3}, the latent heat of evaporation is $\Delta h_v = 120$ kJ kg^{-1}, the heat capacity for constant pressure is the liquid is $c_{p,L} = 1400$ J kg^{-1} K^{-1}, the gravitational acceleration is $g = 9.81$ m s^{-2} and the tank radius is $R = 0.05$ m.

p_0	L_U	ϑ_{sat}	ϑ_{ref}	$\Theta_L = \vartheta_{\text{sat}} - \vartheta_{\text{ref}}$	$\mathbf{R\,Ja}$	D_T
[kPa]	[m]	[K]	[K]	[K]		[m^2 s^{-1}]
220	0.06	327	293	34	35.57	3.76E-8

Table A.53: Modified pressure gradient data for HFE7000 for tests conducted by DAS & HOPFINGER [26] to fit the current excitation of $\eta = 0.78$ by applying the model by DAS & HOPFINGER [26]. Values for δ_t are determined by means of equation (2.91). For known $D_e'^*$ taken from figure 4.36 for known δ_t, u_v is determined by means of equation (2.93), D_e is determined by means of equation (2.87), D_e^* is determined by means of equation (2.94), D_e' is determined by means of equation (2.95).

p_0	t_p	δ_t	$\frac{\partial p}{\partial t}$	$\frac{p_0}{\vartheta_0}\frac{\partial \vartheta_U}{\partial t}$	u_v	D_e	D_e^*	D_e'	$D_e'^*$
[kPa]	[s]	[m]	[Pa s^{-1}]	[Pa s^{-1}]	[m s^{-1}]	[m^2 s^{-1}]		[m^2 s^{-1}]	
220	79	0.021	-621	-280	-1.01E-4	5.58E-8	1.59E-6	1.82E-8	5.21E-7

Table A.54: Properties to predict the pressure gradient for the ESC-B upper stage during ascent phase. The liquid density is $\varrho_L = 64.41$ kg m^{-3} corresponding to an initial pressure of $p_0 = 328$ kPa, the ullage density is $\varrho_U = 4.03$ kg m^{-3}, the latent heat of evaporation is $\Delta h_v = 404.34$ kJ kg^{-1}, the heat capacity for constant pressure is the liquid is $c_{p,L} = 13662.28$ J kg^{-1} K^{-1}, the gravitational acceleration is $g = 9.81$ m s^{-2} and the tank radius is $R = 0.4$ m corresponding to radius of the liquid surface for a fill level of 95%.

p_0	L_U	ϑ_{sat}	ϑ_{ref}	$\Theta_L = \vartheta_{\mathrm{sat}} - \vartheta_{\mathrm{ref}}$	**R Ja**	D_T
[kPa]	[m]	[K]	[K]	[K]		[m^2 s^{-1}]
328	0.14	25	21.5	3.5	1.89	1.18E-7

Table A.55: Modified data to predict the pressure gradient for the ESC-B upper stage during ascent phase fitting the current excitation of $\eta = 0.78$ by applying the model by DAS & HOPFINGER [26]. Values for δ_t are determined by means of equation (2.91). For known $D_e'^*$ taken from figure 4.36 for known δ_t, u_v is determined by means of equation (2.93), D_e is determined by means of equation (2.87), D_e^* is determined by means of equation (2.94), D_e' is determined by means of equation (2.95). For the pressurization (and holding) phase, 1 hour on ground is assumed.

p_0	t_p	δ_t	$\frac{\partial p}{\partial t}$	$\frac{p_0}{\vartheta_0}\frac{\partial \vartheta_U}{\partial t}$	u_v	D_e	D_e^*	D_e'	$D_e'^*$
[kPa]	[s]	[m]	[Pa s^{-1}]	[Pa s^{-1}]	[m s^{-1}]	[m^2 s^{-1}]		[m^2 s^{-1}]	
328	3600	0.061	-1565	-137	-3E-4	1.99E-5	2.51E-5	1.98E-5	2.5E-5

Appendix B

Figures

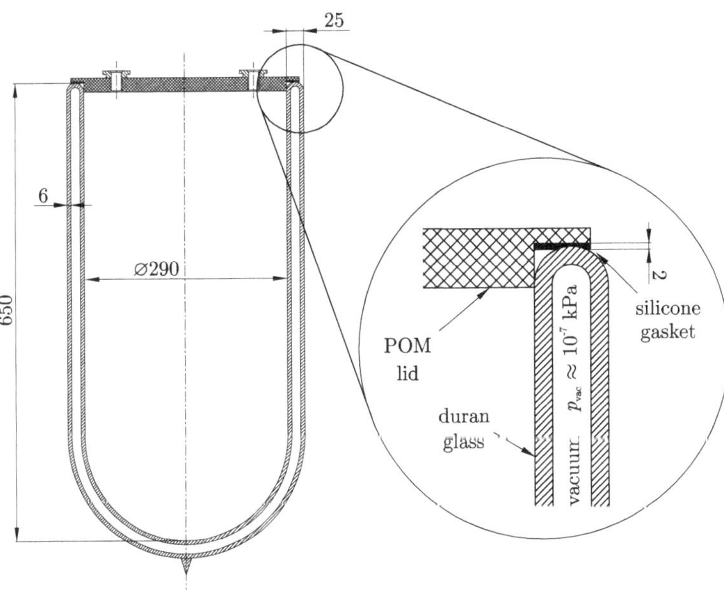

Figure B.1: Drawing of the glass dewar including the main proportions. The detail shows the edge where the tank lid is in contact with the glass dewar.

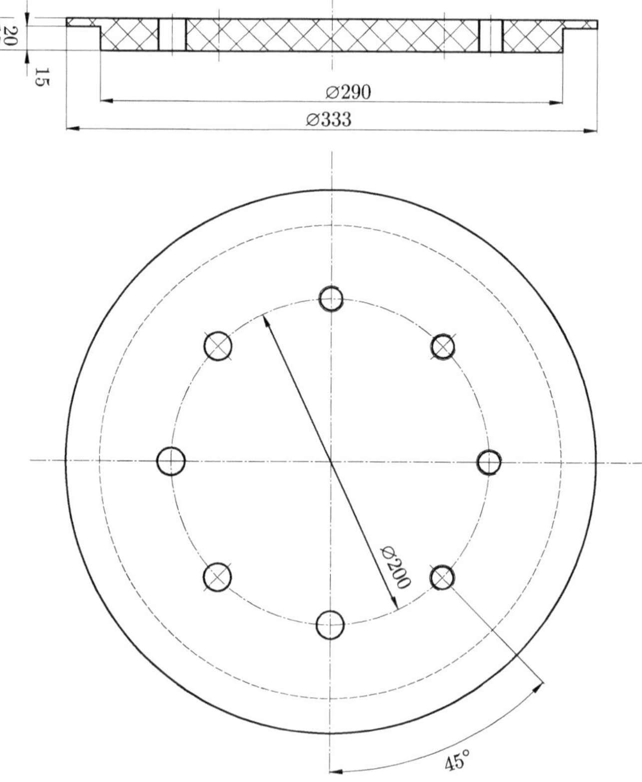

Figure B.2: Drawing of the tank lid that is made of polyacetal. 8 holes are foreseen for connector and supply.

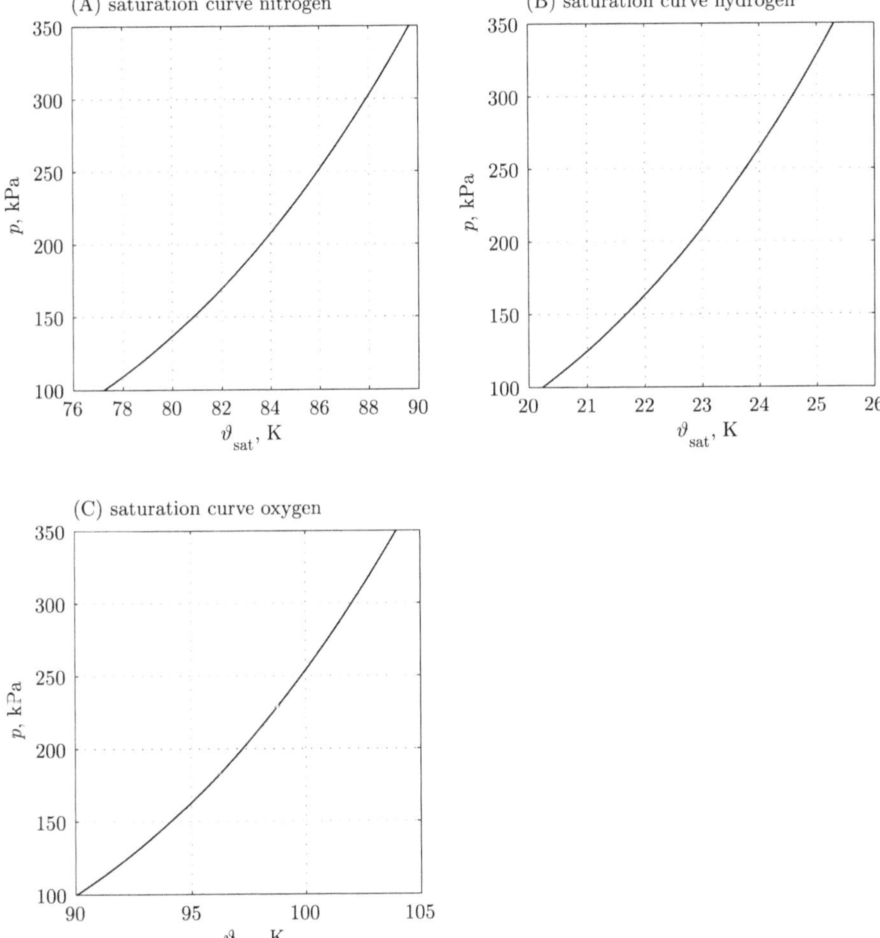

Figure B.3: (A) Saturation curve of liquid nitrogen (LN_2). (B) Saturation curve of liquid hydrogen (LH_2). (C) Saturation curve of liquid oxygen (LOX).

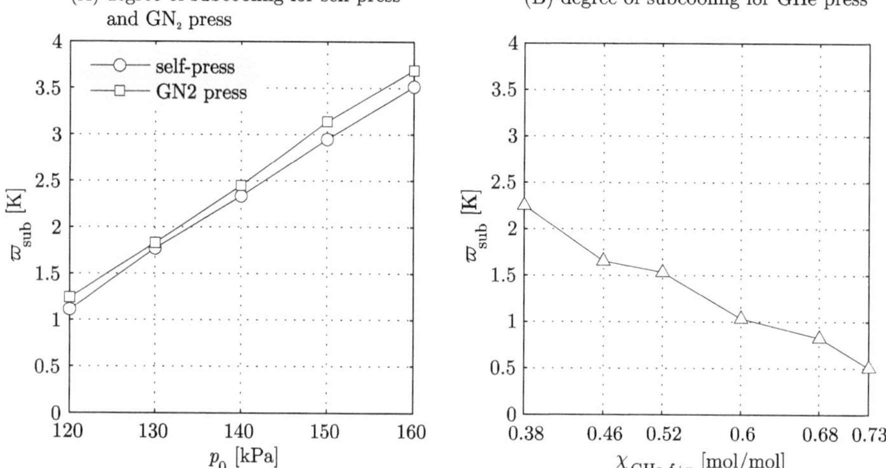

Figure B.4: Degree of subcooling after (A) self-pressurization and GN_2 pressurization as well as (B) GHe pressurization.

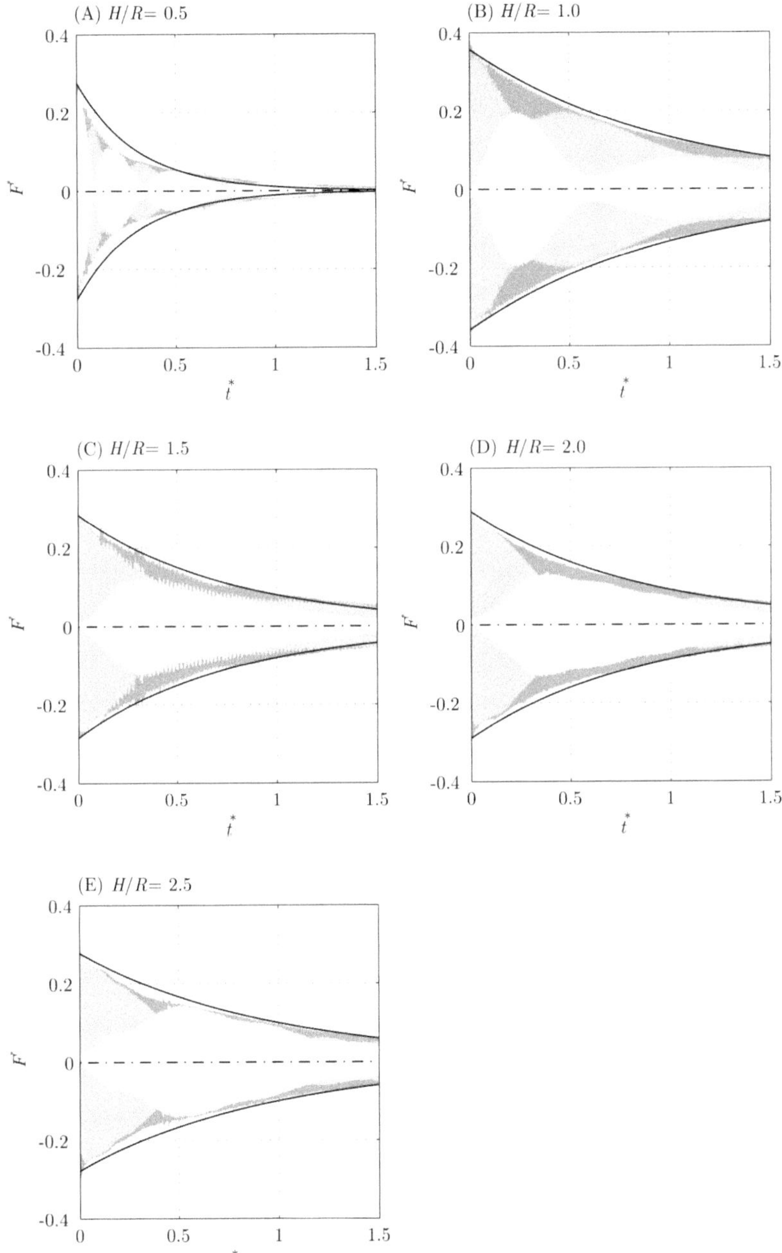

Figure B.5: Exponential decays for (A) $H/R = 0.5$, (B) $H/R = 1.0$, (C) $H/R = 1.5$, (D) $H/R = 2.0$, and (E) $H/R = 2.5$. The light grey zones correspond to F_x while the middle grey zones correspond to F_y. Thus, the dark grey zones correspond to $(F_x^2 + F_y^2)^{1/2}$.